Wetlands - New Perspectives

Edited by Murat Eyvaz and Ahmed Albahnasawi

Published in London, United Kingdom

Wetlands - New Perspectives
http://dx.doi.org/10.5772/intechopen.104315
Edited by Murat Eyvaz and Ahmed Albahnasawi

Contributors
Johan Kieft, Matjaž Glavan, Anisimova Tatiana Yuryevna, Michael Thomas Aide, Samantha Siemers Indi Braden, Kevin Sargent, Miriam Snider, Sven Svenson, Indi Braden, Marissa Wilson, Shakirah Nakasagga, Kariem A. Ghazal, Francisco A. Comín, Murat Eyvaz, Ahmed Albahnasawi

First published in London, United Kingdom, 2023 by IntechOpen
IntechOpen is the global imprint of INTECHOPEN LIMITED, registered in England and Wales, registration number: 11086078, 5 Princes Gate Court, London, SW7 2QJ, United Kingdom
Printed in Croatia

British Library Cataloguing-in-Publication Data
A catalogue record for this book is available from the British Library

Additional hard and PDF copies can be obtained from orders@intechopen.com

Wetlands - New Perspectives
Edited by Murat Eyvaz and Ahmed Albahnasawi
p. cm.

This title is part of the Environmental Sciences Book Series, Volume 7
Topic: Ecosystems and Biodiversity
Series Editor: J. Kevin Summers
Topic Editors: Salustiano Mato, Josefina Garrido and Francisco Ramil

Print ISBN 978-1-83768-894-4
Online ISBN 978-1-83768-895-1
eBook (PDF) ISBN 978-1-83768-896-8
ISSN 2754-6713

IntechOpen Book Series

Environmental Sciences

Volume 7

Aims and Scope of the Series

Scientists have long researched to understand the environment and man's place in it. The search for this knowledge grows in importance as rapid increases in population and economic development intensify humans' stresses on ecosystems. Fortunately, rapid increases in multiple scientific areas are advancing our understanding of environmental sciences. Breakthroughs in computing, molecular biology, ecology, and sustainability science are enhancing our ability to utilize environmental sciences to address real-world problems.

The four topics of this book series - Pollution; Environmental Resilience and Management; Ecosystems and Biodiversity; and Water Science - will address important areas of advancement in the environmental sciences. They will represent an excellent initial grouping of published works on these critical topics.

Meet the Series Editor

J. Kevin Summers is a Senior Research Ecologist at the Environmental Protection Agency's (EPA) Gulf Ecosystem Measurement and Modeling Division. He is currently working with colleagues in the Sustainable and Healthy Communities Program to develop an index of community resilience to natural hazards, an index of human well-being that can be linked to changes in the ecosystem, social and economic services, and a community sustainability tool for communities with populations under 40,000. He leads research efforts for indicator and indices development. Dr. Summers is a systems ecologist and began his career at the EPA in 1989 and has worked in various programs and capacities. This includes leading the National Coastal Assessment in collaboration with the Office of Water which culminated in the award-winning National Coastal Condition Report series (four volumes between 2001 and 2012), and which integrates water quality, sediment quality, habitat, and biological data to assess the ecosystem condition of the United States estuaries. He was acting National Program Director for Ecology for the EPA between 2004 and 2006. He has authored approximately 150 peer-reviewed journal articles, book chapters, and reports and has received many awards for technical accomplishments from the EPA and from outside of the agency. Dr. Summers holds a BA in Zoology and Psychology, an MA in Ecology, and Ph.D. in Systems Ecology/Biology.

Meet the Volume Editors

Dr. Murat Eyvaz is an associate professor in the Environmental Engineering Department, Gebze Technical University, Turkey. His research interests include applications in water and wastewater treatment facilities, electrochemical treatment processes, and filtration systems at the laboratory and pilot scale, membrane processes (forward osmosis, reverse osmosis, membrane bioreactors), membrane manufacturing methods (polymeric membranes, nanofiber membranes, electrospinning), spectrophotometric analyses (UV, atomic absorption spectrophotometry), and chromatographic analyses (gas chromatography, high-pressure liquid chromatography). Dr. Eyvaz has co-authored many journal articles and conference papers and has taken part in many national projects. He serves as an editor for 50 journals and a reviewer for 120 others. He holds four patents on wastewater treatment systems.

Dr. Ahmed Albahnasawi is a research fellow in the Environmental Engineering Department, Gebze Technical University, Turkey. His graduate work focused on the investigation of the treatability of sequential anoxic-aerobic batch reactors followed by ceramic membranes for textile wastewater treatment. Dr. Albahnasawi has published three sci-expanded index journals and participated in three international conferences. His research interests include the application and design of microbial fuel cells integrated with Fenton oxidation for industrial wastewater treatment/solid waste management and monitoring of organic micropollutants by both chromatographic and spectrophotometric analyses.

Contents

Preface

Wetlands are unique ecosystems that provide a wide range of functions and values, from biodiversity conservation to water regulation and purification. They play a crucial role in maintaining the health and well-being of both human societies and the natural environment. The importance of wetlands has been recognized for many years, but in recent times, their significance has become even more prominent due to the growing threats they face, including climate change, habitat loss, and pollution. This book, *Wetlands – New Perspectives*, is a collection of seven chapters that present the latest research and insights on wetlands from different perspectives. It is an outcome of the collaborative effort of several authors who have contributed their knowledge and expertise to provide a comprehensive understanding of wetlands.

The introductory chapter introduces readers to the characteristics, functions, and values of wetlands. This chapter provides a solid foundation for readers to understand the importance of wetlands and their relevance in today's world. Chapter 2 highlights the impacts of climate change and wetland restoration on the water balance components of the coastal wetland. The chapter offers a compelling analysis of the role of wetlands in mitigating climate change and the challenges they face due to changing climatic conditions. Chapter 3 offers a glimpse into the likely status of inland salt lake ecosystems in 2050. The chapter explores the reminiscing and revisiting of Bill Williams and how his contributions can help to predict the future of salt lake ecosystems. Chapter 4 focuses on the economic impacts of the establishment of alternative water retention habitats on agricultural holdings. This chapter provides insights into the economic benefits of using wetlands for water retention and the potential challenges of integrating wetlands into agricultural practices. Chapter 5 presents the challenges and opportunities associated with managing prior converted hydric soils to support agriculture production and maintain ecosystem services. The chapter provides a dedicated outreach to the agriculture community and highlights the potential benefits of sustainable wetland management practices. Chapter 6 explores the monitoring of the properties of an abandoned depleted peat bog to determine the prospects for use. The chapter highlights the potential uses of degraded wetlands and the challenges associated with their restoration and management. Finally, Chapter 7 presents a collaborative approach to the management of fire-resilient peatlands in Indonesia. The chapter highlights the importance of collaboration and community involvement in wetland management and the potential benefits of such an approach.

In summary, this book provides new perspectives on wetlands and highlights the challenges and opportunities associated with their management. It is a valuable resource for wetland managers, researchers, and policymakers who seek to promote the sustainable management of wetlands. We hope that this book will inspire readers to take action to protect and conserve these unique ecosystems for future generations.

Murat Eyvaz
Associate Professor,
Department of Environmental Engineering
Gebze Technical University
Kocaeli, Turkey

Dr. Ahmed Albahnasawi
Department of Environmental Engineering
Gebze Technical University
Kocaeli, Turkey

Chapter 1

Introductory Chapter: Wetlands – Characteristics, Functions, and Values

Murat Eyvaz and Ahmed Albahnasawi

1. Introduction

Wetlands are among the most productive and diverse ecosystems on Earth, supplying a wide range of ecological, social, and economic benefits to humanity [1, 2]. They are defined as areas where the water table is at or near the surface or where the land is covered by water for at least part of the year, resulting in unique hydrological, biogeochemical, and ecological processes [3]. In this chapter, we will provide an overview of the key characteristics, functions, and values of wetlands, as well as their global distribution, threats, and conservation and management challenges.

This book provides a comprehensive overview of wetlands, covering their characteristics, functions, values, threats, and management. It is intended for a broad audience, including students, researchers, practitioners, and policymakers, who are interested in understanding and addressing the challenges and opportunities associated with wetlands. The book is organized into several parts, each of which focuses on a specific aspect of wetlands, such as their physical and biotic components, their role in global and local environmental issues, and their management and restoration strategies.

2. Characteristics of wetlands

Wetlands can be found in a variety of settings, including rivers and lakes, coastal areas, and inland depressions [2]. They are characterized by the presence of water-loving plants or hydrophytes, which have adapted to the wetland environment by developing specialized features, such as floating leaves, waterlogged stems, and oxygen transport systems [3]. These plants supply a variety of ecosystem services, such as nutrient cycling, carbon sequestration, and habitat for wildlife [1].

Wetlands are also distinguished by their hydric soils, which are characterized by saturation or flooding for extended periods and the development of anaerobic conditions that favor the growth of microorganisms, such as bacteria and fungi, that are responsible for decomposing organic matter [2]. These soils are important for regulating the water balance of wetlands, as they can store and release water over time, and also for supporting the growth of wetland vegetation [3].

Another important characteristic of wetlands is their high biodiversity, as they provide habitat for a wide range of plant and animal species, many of which are adapted to the unique wetland environment [4]. Wetlands also serve as important stopovers and breeding sites for migratory birds, such as ducks, geese, and shorebirds, which rely on wetland habitats for food and shelter during their journeys [1].

3. Functions and values of wetlands

Wetlands provide a range of important functions and values to society, including water purification, flood control, carbon sequestration, and recreation [1, 2]. These ecosystem services are essential for human well-being and are often undervalued or overlooked in decision-making processes [4].

Water purification is one of the most important functions of wetlands, as they are able to remove pollutants from water through a variety of physical, chemical, and biological processes [2]. Wetlands act as natural filters, removing sediments, nutrients, and contaminants from water as it flows through them, and also provide a habitat for microorganisms that break down pollutants and improve water quality [3].

Flood control is another important function of wetlands, as they can store and release water over time, reducing the risk of floods downstream. Wetlands act as natural sponges, absorbing excess water during storms and releasing it slowly over time, thereby reducing peak flows and mitigating flood damage [2].

Carbon sequestration is also a vital function of wetlands, as they can store large amounts of carbon in their soils and vegetation [1]. Wetlands are among the most efficient natural carbon sinks, storing up to 30% of the world's soil carbon, despite covering only about 6% of the Earth's land surface [2]. The carbon stored in wetlands is not only important for mitigating climate change but also for maintaining the health and productivity of wetland ecosystems.

In addition to their ecological functions, wetlands also provide a range of social and economic values to society. For example, wetlands are important for recreational activities, such as birdwatching, fishing, and hunting, which generate significant economic benefits for local communities [1]. Wetlands also provide important cultural and spiritual values to indigenous and local communities, who rely on wetlands for subsistence, medicine, and cultural practices [2].

4. Global distribution of wetlands

Wetlands are found all over the world, from the Arctic tundra to the tropics, and from the coasts to the inland areas [3]. However, the distribution of wetlands varies depending on climate, topography, and other environmental factors [4]. The largest wetland complexes are found in the tropics and subtropics, where high rainfall and seasonal flooding create ideal conditions for wetland formation [2].

According to the Ramsar Convention, an international treaty for the conservation and wise use of wetlands, there are over 2300 designated wetland sites around the world, covering an area of over 2.5 million square kilometers (Ramsar [5]). These sites are recognized for their ecological importance and are protected under the convention.

5. Threats to wetlands

Despite their ecological, social, and economic values, wetlands are under threat from a range of human activities, such as land conversion, drainage, pollution, and climate change [1]. Wetland loss and degradation are particularly acute in developing countries, where population growth, poverty, and agricultural expansion are driving the conversion of wetlands for other uses [2].

According to a report by the Ramsar Convention, the world has lost over 35% of its wetlands since the 1970s, with some regions, such as Asia and Europe, experiencing even higher rates of loss (Ramsar [5]). The loss of wetlands has significant ecological and socioeconomic consequences, such as the loss of habitat for wildlife, the reduction of water quality and availability, and the loss of cultural and spiritual values.

6. Conservation and management of wetlands

Conserving and managing wetlands is essential for ensuring their continued provision of ecosystem services and benefits to humanity. There is a range of approaches and tools for wetland conservation and management, including protected areas, restoration, and sustainable use [3].

Protected areas, such as national parks and wildlife reserves, are important for safeguarding wetland ecosystems and their biodiversity. These areas provide legal protection and management frameworks that help prevent the conversion and degradation of wetlands [2].

Wetland restoration is also an important strategy for reversing wetland loss and degradation. Restoration involves the rehabilitation of degraded or destroyed wetlands, through measures such as revegetation, reintroduction of native species, and removal of invasive species [1]. Restoration can help to increase the ecological functioning of wetlands, such as their ability to purify water, store carbon, and provide habitat for wildlife.

Sustainable use of wetlands is another important approach to their conservation and management. This involves balancing the needs of wetland-dependent communities and economic activities with the need to protect the ecological integrity of wetlands [2]. Examples of sustainable use include traditional fishing practices, ecotourism, and sustainable agriculture practices.

In recent years, there has been growing recognition of the importance of integrating traditional ecological knowledge (TEK) into wetland conservation and management [6]. TEK refers to the knowledge, practices, and beliefs of indigenous and local communities about their environment and natural resources. Integrating TEK can help to improve the effectiveness and sustainability of wetland conservation and management by promoting community participation, enhancing the cultural and spiritual values of wetlands, and improving the management of natural resources [6].

7. Conclusion

Wetlands are unique and valuable ecosystems that provide a wide range of ecological, social, and economic services and benefits to humanity. However, wetlands are under threat from a range of human activities, such as land conversion, pollution, and climate change, which are leading to their loss and degradation. Conserving and managing wetlands is essential for ensuring their continued provision of ecosystem services and benefits to humanity. There is a range of approaches and tools for wetland conservation and management, including protected areas, restoration, sustainable use, and integration of traditional ecological knowledge.

Author details

Murat Eyvaz* and Ahmed Albahnasawi
Department of Environmental Engineering, Gebze Technical University, Kocaeli, Turkey

*Address all correspondence to: meyvaz@gtu.edu.tr

References

[1] Davidson NC, Finlayson CM, McInnes RJ. The value of wetlands: importance of scale and landscape setting. Ecological Economics. 2018;**146**:420-429

[2] Mitsch WJ, Gosselink JG. Wetlands. 5th ed. Hoboken, New Jersey, USA: John Wiley & Sons; 2015

[3] Tiner RW. Wetland Indicators: A Guide to Wetland Identification, Delineation, Classification, and Mapping. Boca Raton, Florida, USA: CRC Press; 2017

[4] Zedler JB, Kercher S. Wetland resources: status, trends, ecosystem services, and restorability. Annual Review of Environment and Resources. 2005;**30**:39-74

[5] Ramsar Convention. The Ramsar Convention on Wetlands. 2021 [online] Available at: https://www.ramsar.org/ [Accessed 25 March 2023]

[6] Berkes F, Colding J, Folke C. Rediscovery of traditional ecological knowledge as adaptive management. Ecological Applications. 2000;**10**(5):1251-1262

Chapter 2

The Impacts of Climate Change and Wetland Restoration on the Water Balance Components of the Coastal Wetland

Kariem A. Ghazal

Abstract

The coastal wetlands represent the critical interface between the terrestrial and ocean zones, which have gained vital importance in terms of economic and environmental aspects. Land cover change (LU) and climate change (CC) are considered the determinant factors for the changes in nutrient fluxes, thermal energy, and water balance components (WBCs). These factors are also expected to affect each other through interaction process effects. An essential tool that may be used to evaluate the sustainability and availability of water resources for food security and the ecological health of coastal zones is a hydrological modeling technique. The Heeia coastal wetlands in Hawaii, USA, are used as a case study in this study to evaluate the effects of LU and CC on WBCs.

Keywords: climate change, wetland restoration, SWAT model, water balance, coastal wetland

1. Introduction

Wetlands represent the natural kidney of the coastal environment and the supermarket of unique assemblages of flora and fauna. Wetlands have natural functionalities, which are qualified to be good habitats for birds, aquatic life, plants, and diverse organisms. Therefore, many researchers and policymakers have recently focused on preserving and protecting the wetlands in different regions of the world. For instance, the recent moral and financial support of federal wetlands preservation rules, including "no net loss of wetlands in the United States," has prompted numerous nonprofit organizations to repair the degraded wetlands [1]. In that sense, with assistance from the neighborhood and funding from US environmental protection organizations, the non-profit Hawaii-based organization Kakoo Oiwi has committed to restoring the Heeia coastal wetland (HCW), which is located on the Island of Oahu, Hawaii [2]. Globally, coastal wetlands play an important role against the impacts of climate change (CC), particularly in the coastal zones of pacific Islands such as Hawaii. Hawaiian coastal wetlands provide myriad other benefits associated with protecting coastal communities against storm surges, floods, sea level rise, and CC threats, as well as ecosystem services [3, 4]. Coastal wetlands store and decrease greenhouse emissions through carbon sequestration processes approximately 50% of all carbon is buried in global ocean sediments [5, 6]. As a result, in this chapter, the HCW was used as a case study to demonstrate the importance of wetlands in terms of their vital role in preserving the health environment of coastal regions of the Pacific Islands and mitigating the impacts of CC.

2. Economic and environmental importance

In the Hawaiian Islands, coastal wetlands serve as an important interface between the terrestrial and oceanic zones and are now important for both the ecology and the economy. Coastal wetlands naturally clean water by filtering out sediments and pollutants, converting nutrients, slowing the flow of freshwater from the mountains to the ocean, creating optimal habitats for assemblages of flora and animals, and reducing air temperature during the summer, decreasing greenhouse emissions through carbon sequestration processes, increasing oxygen emission through photosynthesis processes by phytoplankton, kelp, and algal plankton that live in coastal wetland and shoreline of Pacific ocean [7]. Furthermore, the coastal wetlands of Hawaii are regarded as very attractive and productive regions for both tourists and residents [1, 8].

These areas protect Hawaii from flooding, pollution, and the detrimental effects of climatic and land cover changes (LCs). They also operate as sponges, soaking up water during the rainy season and releasing it during the dry season [9, 10]. Many organizations, including scientific research facilities, were forced to take a more proactive approach to preserving and restoring the natural resources of the coastal wetland due to the dynamic nature of these ecosystems. Additionally, the current moral and monetary support for government legislation preserving protected wetlands, such as "no net loss of wetlands in the United States," motivates many non-profit groups to restore the degraded wetlands, such as HCW on Oahu Island [11].

Heeia means "washed way", which is the famous name of Ahupuaʻa, watershed, stream, and fishpond [12]. In the past, the watershed's hydrologic features enabled the indigenous society to meet their food and resource needs from land and sea in a prized coastal region [13]. The Heeia region holds much cultural and historical importance for the people of the Heeia community. The HCW is the southern edge of the Heeia watershed, which was regarded as one of Oahu's most productive coastal areas because of taro and rice farming. Moreover, the Heeia stream estuary, which is located in the area, is thought to be a significant economic resource because it is home to Oahu's largest fishpond. The Heeia watershed (**Figure 1**), which makes up roughly half of the coastal plain, is a steep, mountainous, and narrow valley that eventually converts into a flat marsh zone [8, 14]. Despite its economic and environmental significance, it faces numerous issues related to LC and CC, saltwater intrusion, flooding, the spread of invasive plants, deterioration of the coastal nearshore zone, habitat destruction, and sea level rise (**Figures 1** and **2**) [3–15].

Because of the boundary interaction between the largest federally protected wetland on the island of Oahu, the largest fishpond, and the largest sheltered coral

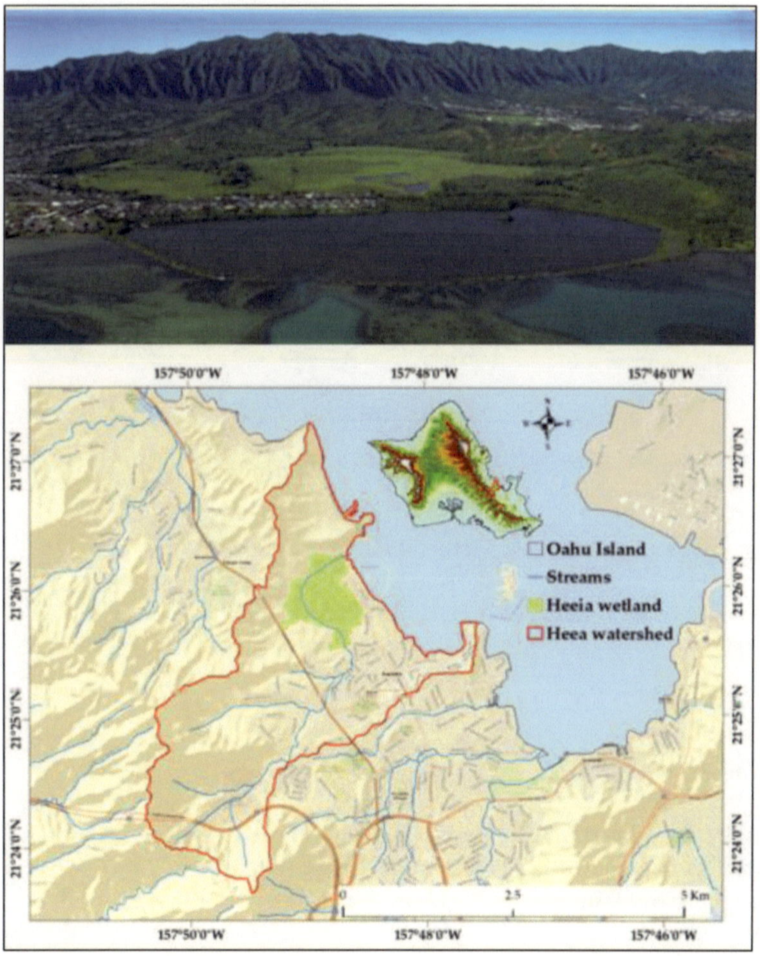

Figure 1.
The landscape view, geographic, and topographic maps of the HCW.

reef system in Kaneohe Bay, the Heeia coastal zone in Hawaii is a typical example of groundwater-dependent ecosystems [16, 17]. In order to protect native ecosystems and marine biodiversity, it is essential to comprehend the processes that take place along the boundary between terrestrial and marine environments [18].

Freshwater flows are critical to the preservation of native adjacent ecosystems. For instance, the availability of nutrients and light influences the growth of diverse groups of plankton in water bodies such as the ocean, lakes, and wetlands. The importance of plankton is obvious because it provides a vital source of food for large aquatic organisms while also reducing greenhouse gas emissions in the coastal zone [7]. Assessing the freshwater discharge and associated nutrient fluxes into the ocean by streams, rivers, and fresh submarine groundwater discharge (FSGD) has piqued the interest of researchers, managers, and policymakers, particularly those concerned with coastal environmental health. [19, 20]. Therefore, to fully comprehend the relationships between coastal hydrological processes and ecosystems, it is necessary to quantify the volumetric freshwater discharge through surface runoff and FSGD in coastal zones.

Figure 2.
The main challenges face HCWs.

3. Wetland restoration

The HCW is a representative example of the Hawaiian wetlands that have been deteriorated and where wetland restoration has been planned [21]. Prior to the 1950s, it was thought to be Oahu Island's most productive environment for both marine and terrestrial food resources [22]. After the 1950s, the Heeia wetland lost the majority of its excellent ecological functions as a result of the invasive California grass (*Urochla mutica*). The degraded marsh cannot be significantly restored using the passive restoration technique (i.e., restoration based on nature's work) unless physical human interventions are directly used in restoration to manage various processes [23]. As a result, human involvement in the restoration of the coastal wetland is crucial for the HCW. In the recently proposed Heeia wetland restoration plan, about 69 hectares of wetland covered in California grass (**Figure 3**) will be converted into organic wetland taro (*Colocasia esculenta*), and eight hectares of wetland mangrove forest will be transformed into wetland sedges papyrus, which will act as a convenient habitat for native birds and a nursery site for young fish [11]. The ecological functioning of a coastal wetland can be improved by wetland restoration initiatives, but the site's hydrologic cycle components may also be significantly impacted. For instance, the wetland evaporates water more quickly than other types of land, reduces air temperature through the evaporation process, traps carbon, maintains stream temperature (by shading, storing, and releasing cool water during dry season), and controls stream flows by acting as a sponge (**Figure 4**) (absorbing water during the wet season and releasing it during the dry season) [25]. Such studies are required to aid the HCW restoration process by assessing the effect of restoration on the hydrologic cycle components. The water balance components (WBCs) of HCW were assessed under current and future LC conditions in this study. The HCW restoration plan was utilized to

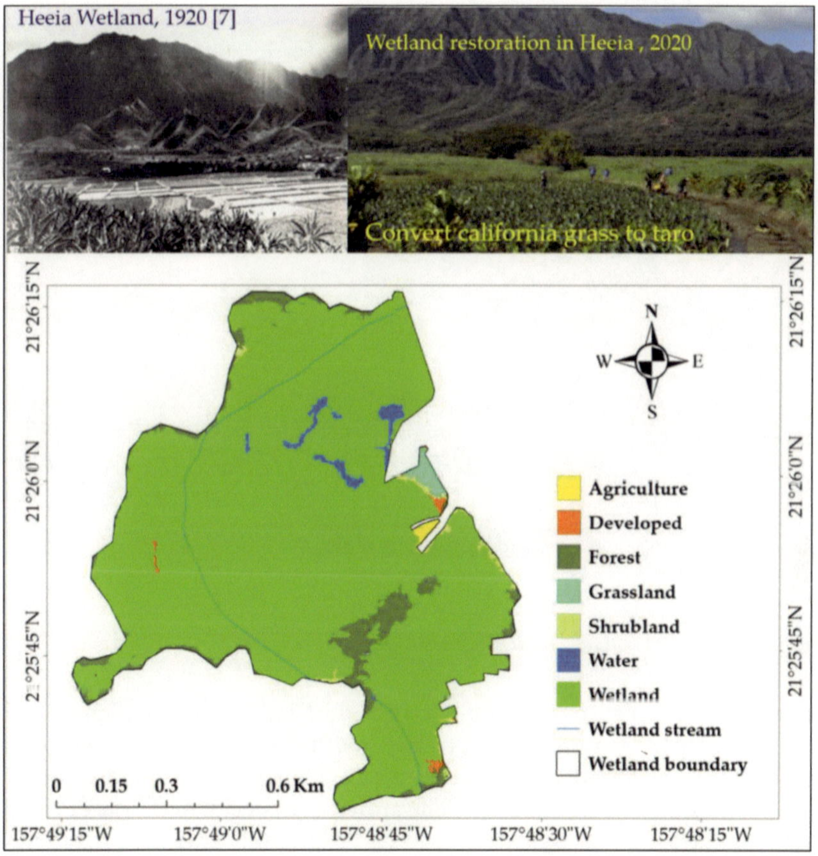

Figure 3.
The pre-development (top, left) and current land use (top, right, and bottom) maps of the Heeia wetland.

develop the future LC [11]. In addition, the study investigated the LC impacts on the spatial and temporal variability of the hydrologic processes within the coastal wetland and its relationship with the hydrologic processes in the highly elevated land of the Heeia watershed [26]. Such studies need a tool to assess the WBCs of HCW.

The Soil and Water Assessment Tool (SWAT) model is a helpful resource for evaluating the WBCs under the conditions of both current and future land use [27]. The SWAT model is suitable for the research area because it is a dynamically processed model, able to adjust the input data of land use and climatic projections over time to predict the future effects of wetland restoration on the WBCs [28]. Additionally, it is computationally efficient to operate at various sizes and appropriate to simulate the consequences of management changes over extended time periods. The model has enormous promise for simulating and analyzing the impact of changing land cover on WBCs [29].

4. The water balance after restoration

Approximately 8% of the Heeia watershed is planned to be converted to taro fields and impoundments. Based on the land use map, the impacts of this change on water

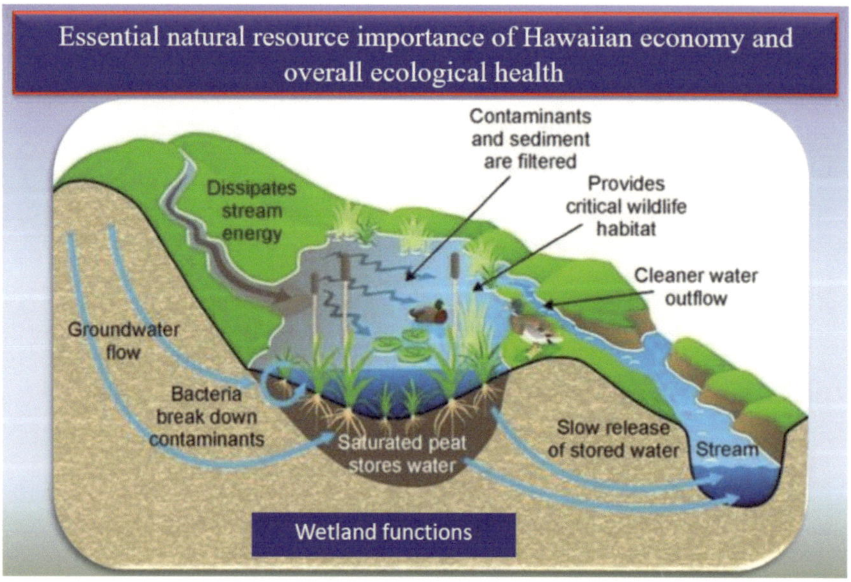

Figure 4.
The hydrological aspects and wetland functions of HCWs [24].

balance were evaluated at three spatial scales of the SWAT model, which included the hydrologic response units (HRUs), subbasins, and watersheds [2]. Within the eight subbasins of the SWAT model in the coastal plain, taro cultivation and a pond were created from the coastal wetland. The anticipated negative effects of changed land cover were depicted in **Figure 5**. Based on the graph, it was anticipated that the restoration would affect the WBCs' yearly average (2002–2014). To maintain ponding water in taro patches, the recharge will be reduced due to soil layer compaction under the taro patches. However, due to lateral seepage from the taro patches, the neighboring areas of the taro patches would receive more recharge [30].

The other elements of the water balance may be affected, and there may be an increase in evaporation from the ponding water area since evapotranspiration (ET) was predicted to grow [4]. Also, as can be predicted, the conversion of an existing wetland (California grass) to taro agriculture would result in a reduction in the site's overall stream flow since stream water would be diverted for taro field irrigation and more pond water would evaporate. A modest percentage change in the restored land cover area relative to the overall watershed area, however, can be blamed for the relatively negligible change in WBCs at the watershed scale. The management of water ponds and taro farming are likely to be to blame for the predicted 41% decline in recharge at the wetland scale under all irrigation diversion scenarios. In comparison, in scenario 4, if 90% of the lowest stream flow was diverted from the main channel, the lateral flow and surface runoff would increase by around 76% and 61%, respectively. While a baseflow reduction of up to 23% is forecast for scenario 4, a substantial increase in surface runoff and lateral flow was predicted to result in a stream flow gain of 13%. Also, it was shown that most WBCs were affected more by the wet season than by the dry season (**Table 1**).

Finally, despite the lack of hydrologic data, the SWAT model accurately captured the temporal variability of the observed daily streamflow hydrographs, exhibiting acceptable performance and satisfactory statistical assessment values. The results

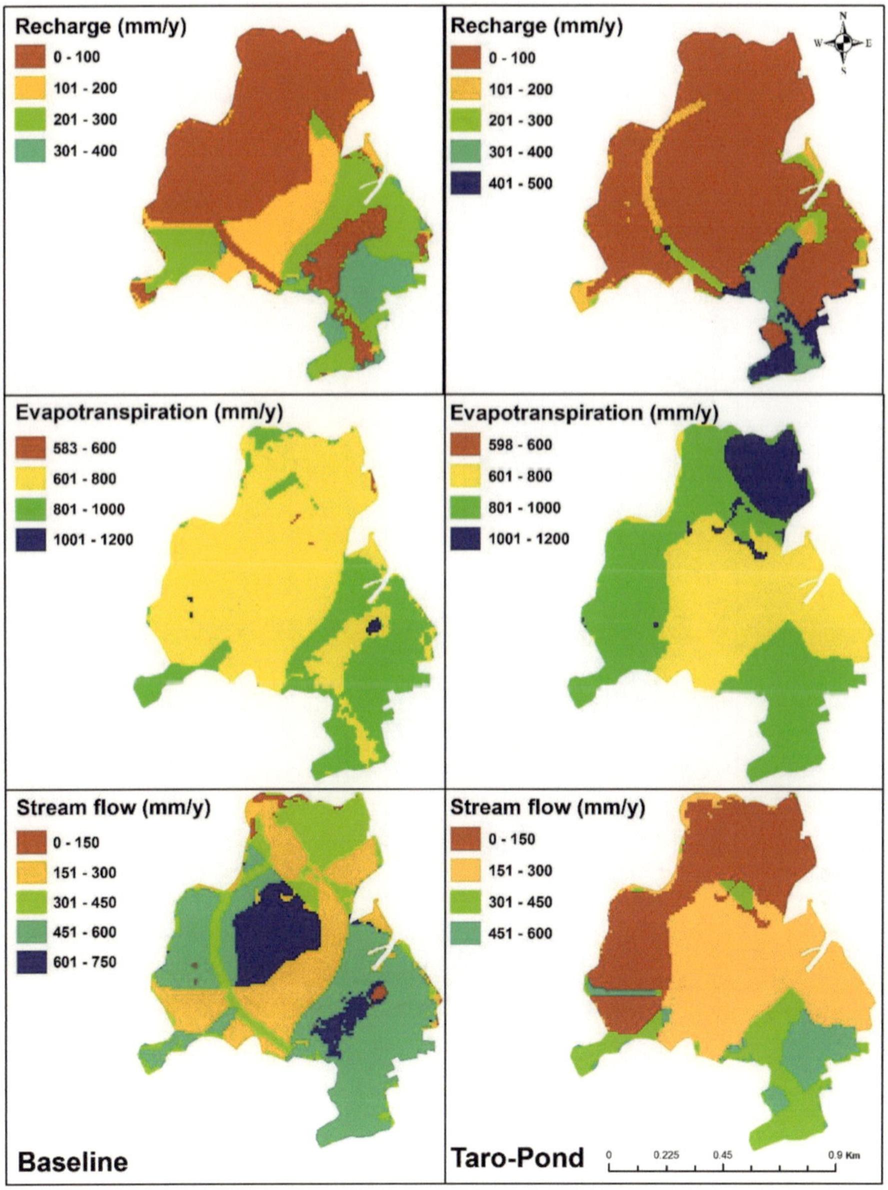

Figure 5.
Yearly average WBCs map of HRUs within the Heeia Wetland [2].

showed that 34% of the watershed's annual rainfall (2043 mm) recharged groundwater (699 mm), 15% of it went to lateral flow (307 mm), 6% of it went to runoff (119 mm), and 45% of it was due to actual evapotranspiration (AET) (917 mm). In addition, 87% of the yearly water supply was contributed by baseflow and lateral flow. In comparison to surface runoff, the baseflow was discovered to be the primary factor in the water yield, as shown in the SWAT output graph (**Figure 6**).

For the wetland area, the HCW restoration plan's effects on WBCs are anticipated to be significant. Furthermore, the restoration strategy is expected to improve lateral flow and surface runoff values while decreasing recharge and baseflow values.

Scale	Scenario	Rainfall	Streamflow	Runoff	LF	BF	Recharge	Soil Moisture	ET	PET
Wetland	Baseline	1065	292	39	91	130	140	115	791	1533
	Irrigation-S1	1065	313	62	137	76	82	144	792	1534
	Irrigation-S2	1065	313	62	137	76	82	144	792	1534
	Irrigation-S3	1065	314	63	138	76	82	144	793	1534
	Irrigation-S4	1065	329	69	147	76	82	147	796	1534
Watershed	Baseline	2043	904	119	306	459	699	171	916	1412
	Irrigation-S1	2043	923	125	331	447	687	176	898	1412
	Irrigation-S2	2043	923	125	331	447	687	176	898	1412
	Irrigation-S3	2043	924	125	331	447	687	176	898	1412
	Irrigation-S4	2043	932	129	336	447	687	177	900	1412

Note: S1 = Scenario one (initial minimum streamflow); S2 = Scenario two (decrease 50% of minimum streamflow); S3 = Scenario three (decrease 75% of minimum streamflow); S4 = Scenario four (decrease 90% of minimum streamflow).
LF = lateral flow; BF = baseflow; ET = evapotranspiration; PET = potential evapotranspiration (except rainfall, all are SWAT outputs).

Table 1.
The percent changes in the seasonal water balance components (WBCs) relative to the baseline for the Heeia Wetland and Watershed.

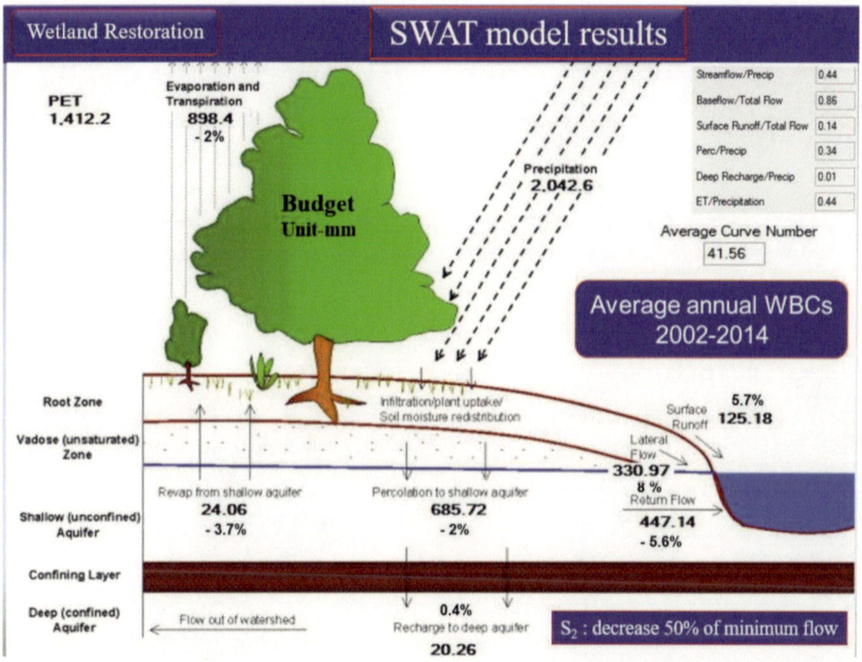

Figure 6.
The WBCs of the Heeia watershed after wetland restoration according to SWAT model outputs [2].

In order to determine the best course of action for achieving sustainable growth of the taro crop without jeopardizing the streamflow values in the main channel and at the downstream fishponds, which are crucial to the downstream coastal ecology of the study area, various irrigation water diversion scenarios were completed to taro fields. According to the results of the study [4], an optimum management strategy for the restoration of the wetland and coastal coastline in the study region is possible by maintaining streamflow and providing the water requirements of the taro patches.

5. The impacts of climate change on water balance components

Both the RCP 4.5 and RCP 8.5 scenarios were evaluated for the relative sensitivity of WBCs to the baseline in terms of percent change for the yearly WBCs due to the combined effects of rainfall, temperature, and solar radiation factors (**Figure 7**). With the exception of PET, both the RCP 4.5 and RCP 8.5 scenarios anticipate a decline in the annual average of WBCs relative to the baseline. The increase in temperature and solar radiation throughout the dry season is anticipated to result in a continuous rise in the relative percent change of PET. The AET did, however, fall short of the baseline value, most likely as a result of a decline in rainfall that constrained the availability of soil moisture. Rainfall was therefore identified as the determining element [31, 32]. The effect of rainfall at the coastal region change would be more pronounced at the coastal region, compared to the upstream regions. Because of the fluctuating rainfall, temperature, and solar radiation under both scenarios (RCP4.5 and 8.5) of CC, the results using monthly time steps showed that the dry season generated a more severe relative negative shift in the WBCs than the rainy

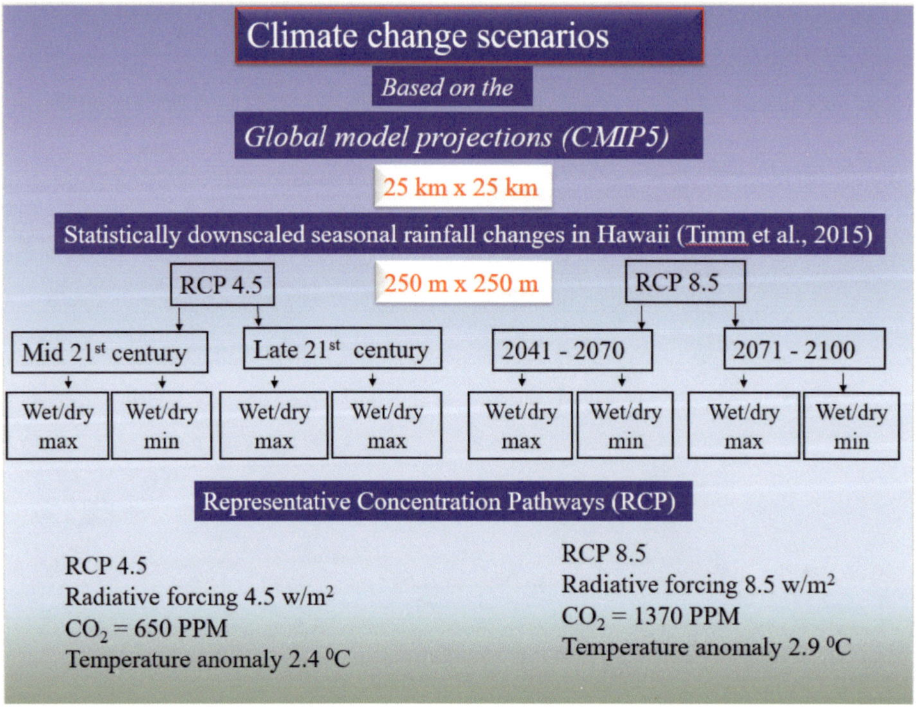

Figure 7.
The CC scenarios of Hawaii Islands [4].

season [4]. The relative negative change in WBCs was larger in the coastal wetland than further upland in the watershed due to the variance in climatic conditions at both the geographical and temporal scales [33]. Moreover, RCP 8.5 had a greater relative negative change in the dry season than RCP 4.5, particularly for the late (2080s) period as compared to the middle (2050s) period. Due to climatic parameter variation, these adverse effects were more obvious for the seasonal changes in recharge, surface runoff, lateral flow, and rainfall, especially at the wetland scale as compared to the entire watershed scale [32, 34, 35]. Due to the low value of recharge within the wetland, there was a large value of relative change in recharge compared to other components. The results showed that streamflow dropped, especially during the late 2080s of RCP 8.5 (**Figures 8** and **9**). In addition, the CC is expected to cause decrease in the streamflow, baseflow, and groundwater recharge for the whole watershed (**Figure 10**). This could be due to a consistent decrease in rainfall for both wet and dry seasons.

6. Conclusion

The HCW restoration is significantly influenced by the hydrological processes of the whole watershed. In order to prioritize the actions of the coastal wetland restoration, it is important to examine the hydrological processes at the watershed scale and comprehend their influences on the coastal wetland. Additionally, it is believed that managing the water resources of coastal wetlands is the key to maximizing

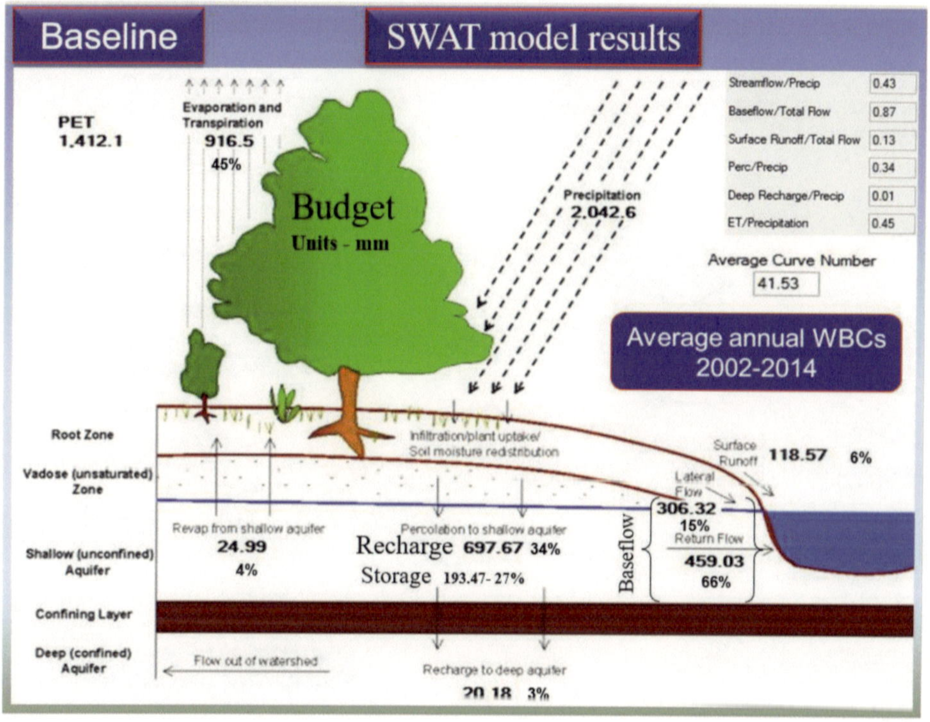

Figure 8.
The baseline of WBCs of Heeia watershed [4].

the sustainability of the coastal ecosystems. Tools that can assist in evaluating the coastal water resources are required for such an approach. Hydrological models were the tools utilized to evaluate the management of the water resources in the Heeia coastal zone.

The coastal wetland restoration would be expected to be impacted by the WBCs. When compared to the baseline, the ET is expected to rise, potentially reducing the other WBCs and increasing the ponding water area. Reduced baseflow would lead to a decrease in stream flow overall as a result of the conversion of an existing wetland (California grass) to taro agriculture. When water diversion was adjusted to 50%, 75%, and 90% of the minimum streamflow, the effects of applied irrigation diversions were roughly 23, 109, 437, and 3886 mm/y, relative to the baseline (no-irrigation), after the restoration of taro farming and the construction of ponds. The minor percent change in California grassland area relative to the Watershed's area may be the cause of the generally negligible change in WBCs at the Watershed scale. The WBCs at the wetland scale, however, were considerably impacted by this land cover shift. In contrast to ET, surface runoff, and lateral flow, for instance, recharge is projected to increase.

The combined effects of wetland restoration and CC may have a substantial impact on the WBCs of Heeia Wetland. The variance in rainfall over both space and time was the main contributor to the adverse effect on WBCs. The components that were most vulnerable to the combined effects of land cover and climatic changes, particularly during the dry season, were recharge and baseflow. The WBCs were generally more impacted in the late 2080s than in the 2050s timeframe.

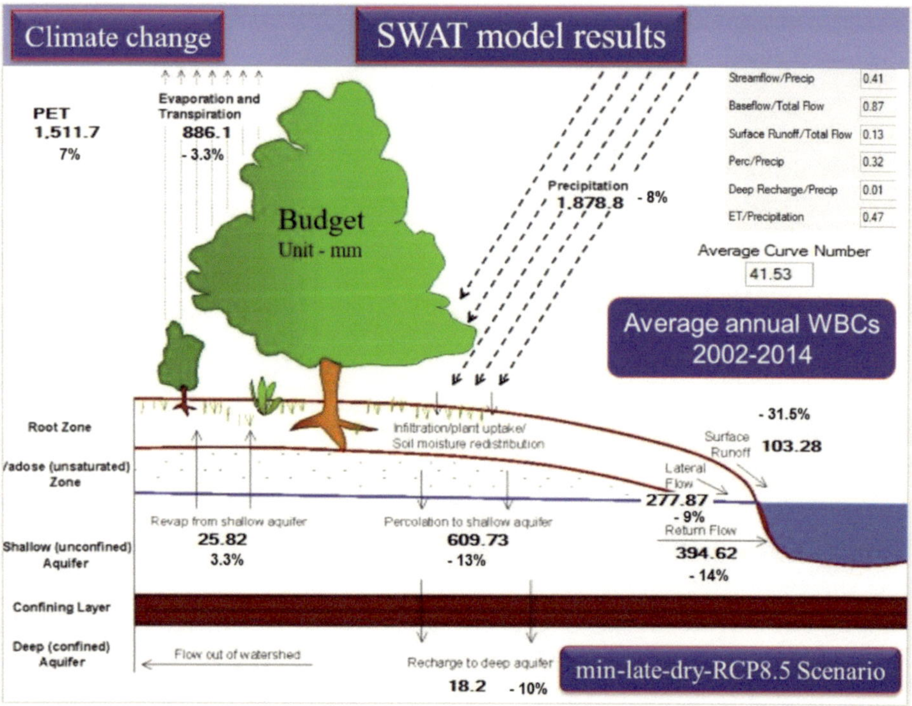

Figure 9.
The CC impacts on WBCs of Heeia watershed [4].

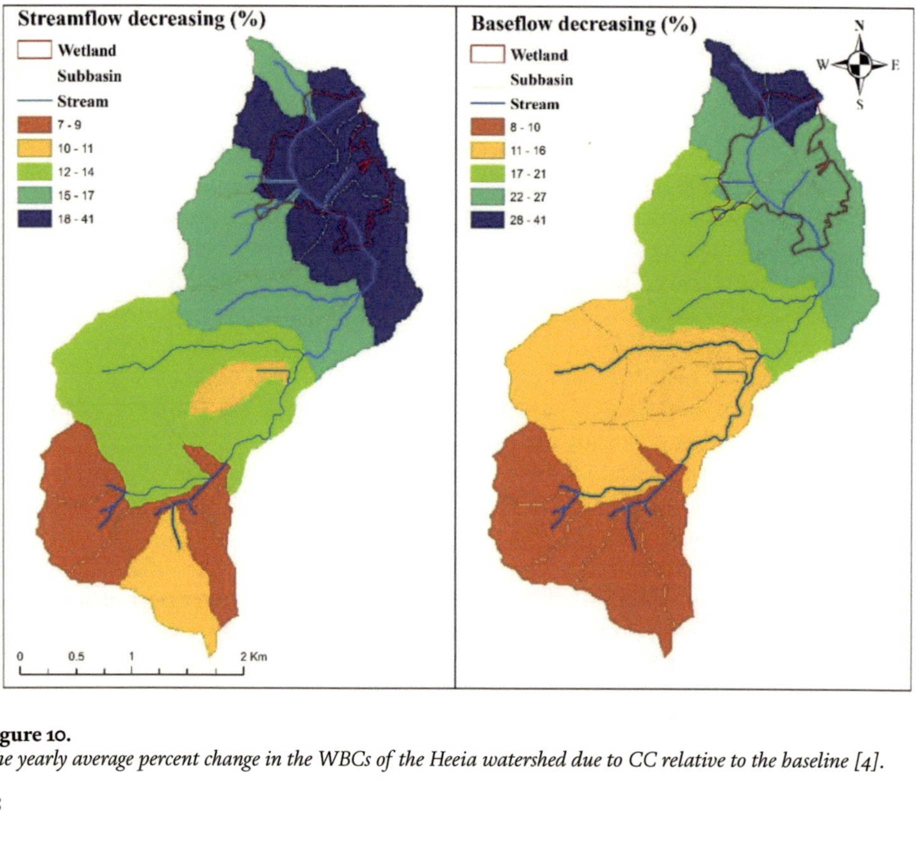

Figure 10.
The yearly average percent change in the WBCs of the Heeia watershed due to CC relative to the baseline [4].

Acknowledgements

The author thanks the Kāko'o 'Ōiwi community for facilitating the research and the Iraqi Ministry of Higher Education and Scientific Research funded the author's studies at the University of Hawaii at Manoa. The author expresses gratitude to both groups. This is UNIHI-SEAGRANT-JC-14-65, publication number 10699 of the School of Ocean and Earth Science and Technology (SOEST), and WRRC-CP-2019-09 of the Water Resources Research Center (WRRC), University of Hawaii at Manoa, Honolulu, Hawaii. The author also acknowledges Dr. Mohammad H. Dawood, a lecturer and researcher at the University of Kufa's Faculty of Agriculture, for his insightful remarks made throughout the manuscript.

Conflict of interest

The author declares no conflict of interest.

Author details

Kariem A. Ghazal
Department of Soil and Water Sciences, College of Agriculture, University of Kufa, El-Najaf, Iraq

*Address all correspondence to: kareem.alzubaidi@uokufa.edu.iq

References

[1] Mitsch WJ, Gosselink JG. Wetlands. John Wiley & Sons; 2015

[2] Ghazal KA, Leta OT, El-Kadi AI, Dulai H. Impact of coastal wetland restoration plan on the water balance components of Heeia Watershed, Hawaii. Hydrology. 2020;**7**(4):86

[3] Rotzoll K, Fletcher CH. Assessment of groundwater inundation as a consequence of sea-level rise. Nature Climate Change. 2013;**3**(5):477-481

[4] Ghazal KA, Leta OT, El-Kadi AI, Dulai H. Assessment of wetland restoration and climate change impacts on water balance components of the Heeia Coastal wetland in Hawaii. Hydrology. 2019;**6**(2):37

[5] Pörtner H-O, Roberts DC, Adams H, Adler C, Aldunce P, Ali E, et al. Climate change 2022: impacts, adaptation and vulnerability. IPCC Sixth Assessment Report. 2022

[6] Lovelock CE, Fourqurean JW, Morris JT. Modeled CO_2 emissions from coastal wetland transitions to other land uses: tidal marshes, mangrove forests, and seagrass beds. Frontiers in Marine Science. 2017;**4**:143

[7] Libes S. Introduction to Marine Biogeochemistry. Academic Press. 2011. ISBN: 9780080916644

[8] Izuka SK, Hill BR, Shade PJ, Tribble GW. Geohydrology and possible transport routes of polychlorinated biphenyls in Haiku Valley, Oahu, Hawaii: US Department of the Interior, US Geological Survey; 1993

[9] Bruland G. Coastal wetlands: function and role in reducing impact of land-based management. Coastal watershed management. 2008;**13**:85

[10] Bantilan-Smith M, Bruland GL, MacKenzie RA, Henry AR, Ryder CR. A comparison of the vegetation and soils of natural, restored, and created coastal lowland wetlands in Hawai 'i. Wetlands. 2009;**29**(3):1023-1035

[11] Kakoo Oiwi. Heeia Wetlands Restoration. 2011. Contract No.: POH-2010-00159

[12] Devaney DM. Kaneohe: A History of Change, 1778-1950 ICON Group International; 1976

[13] Hunter CL, Evans CW. Coral reefs in Kaneohe Bay, Hawaii: two centuries of western influence and two decades of data. Bulletin of Marine Science. 1995;**57**(2):501-515

[14] Kakoo Oiwi. Heeia Wetland Restoration Strategic Plan 2010-2015. 2010

[15] Smith SV, Kimmerer WJ, Laws EA, Brock RE, Walsh TW. Kaneohe Bay sewage diversion experiment: perspectives on ecosystem responses to nutritional perturbation. University of Hawaii Press. 1981;**35(**4)279-395

[16] Jokiel PL. Jokiel's illustrated scientific guide to Kaneohe Bay, Oahu. Hawaiian Coral Reef Assessment and Monitoring Program, Hawaii Institute of Marine Biology, Kaneohe, Hawaii; 1991

[17] Dulai H, Kleven A, Ruttenberg K, Briggs R, Thomas F. Evaluation of submarine groundwater discharge as a coastal nutrient source and its role in coastal groundwater quality and quantity. In: Fares A, editor.

Emerging Issues in Groundwater Resources. Advance in Water Security. Springer; 2016:187-221. ISSN: 2523-3572

[18] Wilder RJ, Tegner MJ, Dayton PK. Saving marine biodiversity. Issues in Science and Technology. 1999;**15**(3):57-64

[19] Ghazal KA, Leta OT, El-Kadi AI, Dulai H. Quantifying Dissolved Silicate Fluxes across Heeia Shoreline in Hawaii Via Integrated Hydrological Modeling Approach. Journal of Geosciences. 2018;**6**:147-156

[20] Crossland CJ, Kremer HH, Lindeboom H, Crossland JIM, Le Tissier MD. Coastal Fluxes in the Anthropocene: The Land-Ocean Interactions in the Coastal Zone Project of the International Geosphere-Biosphere Programme. Springer Science & Business Media; 2005

[21] Henry AR. Strategic plan for wetland conservation in Hawai`I. Pacific Coast Joint Venture Hawai`I; 2006

[22] Kailua Bay Advisory Council. Ko'olaupoko Watershed Restoration Action Strategy Kailua Bay Advisory Council (KBAC). Hawaii state: Hawaii's Department of Health; 2007 Contract No.: ASO Log No. 05-080

[23] Kusler J, Kentula M. Wetland creation and restoration: the status of the science. v. 1. Regional reviews--v. 2. Perspectives. EPA-600/3-(USA) no 89/038a-b. 1989

[24] Kentula ME. Foreword: monitoring wetlands at the watershed scale. Wetlands. 2007;**27**(3):412-415

[25] Bullock A, Acreman M. The role of wetlands in the hydrological cycle. Hydrology and Earth System Sciences. 2003;**7**(3):358-389

[26] Cortes G, Ragettli S, Pellicciotti F, McPhee J, editors. Hydrological models and data scarcity: on the quest for a model structure appropriate for modeling water availability under the present and future climate. AGU Fall Meeting Abstracts; 2011

[27] Green C, Tomer M, Di Luzio M, Arnold J. Hydrologic evaluation of the soil and water assessment tool for a large tile-drained watershed in Iowa. Transactions of the ASAE. Scientific Research Publisher. 2006;**49**(2):413-422

[28] Arnold JG, Srinivasan R, Muttiah RS, Williams JR. Large Area Hydrologic Modeling and Assessment part I: Model Development 1. Wiley Online Library; 1998. DOI: 10.4172/2157-7587.1000216, ISSN: 2157-7587 HYCR

[29] Kiros G, Shetty A, Nandagiri L. Performance evaluation of SWAT model for land use and land cover changes in semi-arid climatic conditions: a review. Hydrology: Current Research. 2015;**2015**

[30] Xie X, Cui Y. Development and test of SWAT for modeling hydrological processes in irrigation districts with paddy rice. Journal of Hydrology. 2011;**396**(1):61-71

[31] Erwin KL. Wetlands and global climate change: the role of wetland restoration in a changing world. Wetlands Ecology and Management. 2009;**17**(1):71-84

[32] Safeeq M, Fares A. Hydrologic response of a Hawaiian watershed to future climate change scenarios. Hydrological Processes. 2012;**26**(18):2745-2764

[33] Giambelluca T, Chen Q, Frazier A, Price JP, Chen Y-L, Chu P-S, et al. The Rainfall Atlas of Hawai'i. 2011

[34] Timm OE, Giambelluca TW, Diaz HF. Statistical downscaling of rainfall changes in Hawai ʻi based on the CMIP5 global model projections. Journal of Geophysical Research: Atmospheres. 2015;**120**(1):92-112

[35] Diaz HF, Giambelluca TW, Eischeid JK. Changes in the vertical profiles of mean temperature and humidity in the Hawaiian Islands. Global and Planetary Change. 2011;77(1-2):21-25

Chapter 3

The Likely Status of Inland Salt Lake Ecosystems in 2050: Reminiscing and Revisiting Bill Williams

Francisco A. Comín

Abstract

The classical management approach of inland saline lake ecosystems focused on ecological issues, including conserving their biological communities and physical-chemical characteristics. However, the peculiarity of saline lake ecosystems is that they are in a limited watershed, in many cases a closed watershed. So, its management should be planned and performed at watershed scale, which has been frequently neglected. W.D. (Bill) Williams was one of the key persons rising awareness for conservation and promoting their rationale management based on scientific research results. This work shows, through a literature review, that classical management approaches included returning impacted salt lakes to initial conditions through, mostly, eliminating the processes impacting them. At the turn of the century, a wider approach emerged. In addition to focusing on watershed scale management, the integration of social, economic, and environmental issues was incorporated into management proposals by different authors. Lake Gallocanta case study is described and discussed as a paradigm of inland salt lake management. The status of inland salt lakes will improve in the future if land cover reparcelling, and rationale uses of water in the watershed are incorporated, considering adaptive practices to climate change impacts and a balanced provision of ecosystem services.

Keywords: management, watershed, climate change, fluctuations, inland salt lakes

1. Introduction

The conservation of inland salt lakes (ISL) has been a matter of interest for the last four decades [1, 2]. Although scientific aspects of ISL have been investigated since long ago [3, 4], their management focused on the extraction of minerals for a long time during the last century and before [5, 6]. The improvement of their knowledge of ecosystems [7] increased the interest for their conservation since the last decades of the twentieth century [8, 9]. The triennial conference series on inland salt lakes started in Australia [10] and followed under the auspices of the International Society for Salt Lake Research was instrumental in extending the interest and knowledge on inland salt lakes. William (Bill) D. Williams early realized the relevance of IS L as a unique ecosystem and encouraged both their study from the scientific point of view and their conservation in a good ecological state [11, 12]. All around the world, many others contributed to increasing the information on the biological components, comprehending the hydro-geochemical processes regulating ISL dynamics, and the interest in their conservation [13, 14].

By the end of the last century, most interests was devoted to classical issues as species conservation and hydrology-related impacts. The desiccation of ISL because of human land uses in their watershed, called anthropogenic drought, was a challenge for their conservation. Hydrological modeling was used as a tool to show the relationships between water flows in their watersheds and water level changes in the ISL. Many detailed studies were dedicated to knowing the trophic webs and their dynamics in relation to chemical characteristics of the water. However, their conservation and management received much less attention by the scientific community.

Early this century, several authors claimed the global trend of degradation affecting lakes [15, 16]. This reflected the previous alarm calls that arose by scientists all around the world to be aware of the causes of the degradation of lakes [17]. ISL were one of those most negatively affected by desiccation due to water diversion and abstraction in their watersheds. Other negative impacts included pollution, salinization, direct occupation of the lake's shores by urban developments and mining activities. Also, global climate change has been reported as a major factor causing a general decrease in the water level of many lakes [18].

In some way, these papers reflected the ideas of others who described the degradation of salt lakes globally during the last decades of the twentieth century and anticipated a deficient ecological state for the first decades of the twenty-first century [19, 20]. During the last two decades, concern about the state of ISL increased simultaneously with the awareness of the environment in general after the Rio Declaration on Environment and Development and the Sustainable Development Declaration of the United Nations in 1992 and 2015, respectively, and the successive reports of the International Panel for Climate Change, all of them demonstrating the relationships between human activities and degradation of the environment and suggesting new models of development based on the preservation of natural resources.

This work follows a similar trail as in the above-mentioned papers to discuss the likely status of salt lakes by the mid-twenty-first century. The projection to 2050 is an ideal figure just to define a period long enough to observe changes in the status of salt lakes. A literature review is done to ascertain major topics of interest of authors with respect to the conservation and management of salt lakes. Then, following Bill's footprints, the likely status of salt lakes is discussed after a driver-pressure-state-impact-response route. Finally, new paradigms for the management of salt lakes are presented using a few well-known lakes as case studies. Here, we focus on athalassic inland salt lakes, although a few inland thalassic lakes are quoted as examples of the specific issue. Coastal lagoons are not considered here, as they are driven by processes linked to coastal and marine hydrodynamics, which are quite different from those regulating inland salt lakes.

2. Literature review

Vosviewer, a visualization software [21], was applied to a series of 299 documents (papers published in scientific journals and proceedings of congresses) obtained from Web of Science using the searching terms "salt lakes", "conservation" and

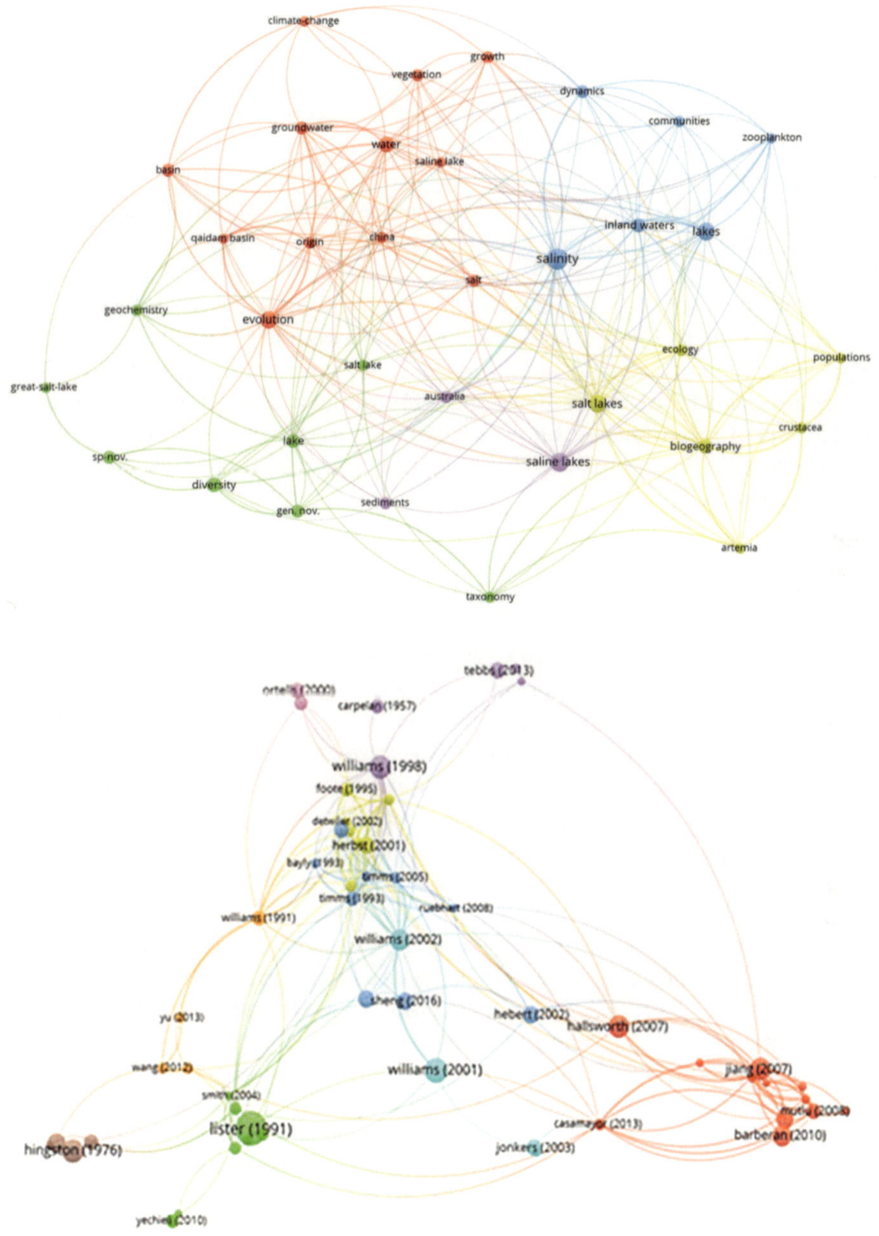

Figure 1.
Visualization maps generated by Vosviewer showing: Six major clusters of topics of interest in the study of inland salt lakes after the literature review (above); relationships between authors and papers published by most cited authors (below). Each cluster is represented with a different color; the font size of the nodes in the graph represents the relative frequency of occurrence or both topics, above and authors, below.

"management" on 15 July 2021 for the period 1992–2020 after removing from the initial list of 432 documents obtained those not related to inland salt lakes.

The revision of scientific publications showed that just a few words were most frequently used to refer the topics of interest: salinity (and its related words plants, water, diversity, growth, wetland and salt), saline and salt lakes (and its related words biogeography, zooplankton, crustacea, tolerance and biomass), evolution (and its related words basin, origin, groundwater, system and geochemistry), inland waters (and its related words shallow, climate change and remote sensing). A few other words – taxonomy, genus nov., sp. nov., prokaryotic diversity, gradient – were also less frequently used as representative of topics of interest in papers on inland salt lakes. Many other words were much less frequently used to simplify the relevant subjects of scientific papers dealing with inland salt lakes (e.g. nutrient, salt tolerance, eutrophication, branchiopoda, habitats, identification, desert, archaea and hypersaline environments).

These six groups of words corresponding to topics of interest in the study of ISL are represented in the Vosviewer visualization map in **Figure 1**. It is remarkable that the most outstanding author in the set of publications selected with this literature review is (Bill) Williams (**Figure 1**), pioneer in the study of inland salt lakes who was promoting their knowledge and conservation all around the world for many years. Following his wake, many other authors later reviewed the state of inland salt lakes.

3. The conservation of inland salt lakes and their status in 2025

The driver-pressure-state-impact-response (DPSIR) approach is a causal framework to describe the interactions between society and the environment developed by the European Environment Agency [22]. Data and information on the five steps are collected from the scientific literature, official conservation departments and Internetlinked. Here, the DPSIR analysis is applied generically for ISL in general (**Table 1**). A special emphasis is devoted to lakes previously reviewed by Bill Williams [20] and other authors to comment on their management progress and their likely status in 2050.

Burning fossil fuels, which is the major cause of global climate change, is related to the decrease of the water level of many lakes and, consequently, increased desiccated periods (both frequency and duration) and related impacts as increased penetration of ultraviolet radiation. A few lakes in Pakistan are examples of this case, with reduced habitat representativeness and biodiversity and altered life cycles of aquatic species [23]. Many other ISL have been referred to with decreasing water levels related to climate change [18, 24–26]. The relationship between climate change and inland water shrinking is common in many lakes around the world [27, 28]. However, a detailed water budget is needed to discern the relative contributions of climate variability and human impacts on lake inflows [14].

Land cover and land use changes in the watershed are major drivers of changes in the state of ISL. Their pressures include alteration of surface and groundwater flows and, consequently, affect the water cycle and impact both the quantity and the quality of the water in the lakes, as well as alter its salinity and habitats. The expansion of irrigated agriculture is the origin of decreased water levels, water quality degradation and biodiversity loss in many lakes around the world. Aral, Mono, Dead Sea, Qinghai, Corangamite, Winnemuca and Owens are ISL submitted to these impacts caused by changes in land cover and land use changes in their watersheds with changes in water uses.

The Aral Sea has been one of the most cited closed lakes impacted by diversions of surface freshwater inputs to irrigate new developments of agriculture, which caused a catastrophic water level decrease and took the lake state to collapse. In contrast,

Drivers	Pressure	State	Impact	Case studies
Global use of fossil fuels & land cover & land use changes,	Climate change	increased penetration of ultraviolet radiation, Decreased water level, increased desiccated periods: temporary salt lakes in these regions will remain drier for longer, and permanent salt lakes will become smaller and more saline	Reduced biodiversity, altered life cycles of species. Lack of habitat representativeness	Ucchali, Khabbeki, Jahlar Lakes in Salt Range (Pakistan).
Expansion of irrigated agriculture & other water uses in the watershed (urban wastewater discharges)	Surface inflow diversion	Altered water cycle Reduction of characteristic habitats.	Lowered water level. Increased salinity. Water hyper-salinization	Aral, Mono, Pyramid, Dead Sea Qinghai, Corangamite, Winnemuca, Owens
Agriculture irrigation by groundwater pumping	Surface & groundwater Inflows of salinized water	pollution	increased salinity Altered habitats	Lake Tulare Lakes in Salt Range (Pakistan)
Clearance of the natural vegetation and other land use changes within catchments	Increased sediment loads Altered runoffs	Increased salinity^ Altered water cycle A more halotolerant community one	Secondary (anthropogenic) salinization Replacement of biological communities Decreased biodiversity	Lake Pyramid
Urban development in the catchment	Loss of lake area and habitats	Decreased esthetic, scientific values Pollution	Loss of lake area	Great Sal Lake
Mining (for mineral extraction)	Extraction of mineral and biotic components Soil-water interactions	Loss of water quality	Pollution	Chaidamu Basin, Qinghai province lakes of China, Zabuye Salt Lake, Jiezhaka Salt Lake and Longmucuo Salt Lakes in Ali region of Tibet.
Exotic species	Biological disturbances	Altered trophic structure	Elimination of species, changed community	Rottnest Island and Lake Hayward// Caspian Sea

Table 1.
The DPSIR approach applied to ISL reminiscing some of the lakes discussed by bill Williams.

the water level of the Caspian Sea, the largest in the world endorheic saline lake, has steadily increased. These two lakes are just 500 km far from each other. However, they contrast with respect to water management in the watershed. The withdrawal for irrigation is 90% of the water resources of the Aral watershed, which occupies a semiarid territory, and just 10% of the water resources in the watershed of the Caspian Sea, which occupies a temperate territory [29].

Clearance of the natural vegetation is another driver who causes increased sediment loads and altered runoffs in ISL. A management strategy is being developed in the Truckee River Watershed (Nevada, USA) to reduce sediment loads and improve the water quality of the water inflowing Lake Pyramid. It includes the restoration of riparian, aquatic and wetland habitats and planning new urban and rural land cover changes to avoid easily erodible areas [30].

Urban development in the lake shores and watershed is a driver causing decreased lake area. This is also a relevant driver-pressure factor affecting Great Salt Lake since the early 1900s. Diversion of water inflows, constructed infrastructure in the lake, and dust storms from desiccated areas are also key pressures not only for the lake state but also for the human population living near the lake [31]. Similar actions dominated the degradation of Salton Sea (California, USA), an artificially created ISL over old marine deposits which receives huge amounts of pollutants (nutrients, pesticides) used in agricultural activities in its watershed [32].

Mining is also a major driver in those ISL where mineral extraction, mostly linked to evaporative deposits, takes place. Extraction of salt deposited in the lakebed to obtain salt itself or metals associated with the salt (e.g. lithium) causes loss of the water quality. This is the case for the lakes located in the Qaidam Basin in Tibet and Qinghai Basin in China and in the Andean's plateau [33, 34].

Invasive alien species trigger biological disturbances and alter lakes' trophic structure, as it happened in Caspian and Aral Seas after the introduction of fish for commercial purposes [27]. The consequence is a changed biological community, including the disappearance of key species. In many cases, the introduction of alien species was not successful because they did not adapt to the environmental conditions of salt lakes. However, some species were able to acclimatize and had significant impacts on native populations. This is a major aspect to consider while trying to manipulate the water level and inland salt lakes' water salinity.

The conclusion is that there are big differences between ISL with respect to their ecological states. In those countries where their natural resources are highly appreciated, the status is good or, at least, there is a major concern to preserve all their values and to establish regulations to combine a rational use of their resources. In Europe, where ISL are relatively small, there are European regulations to preserve their biological populations and their habitats. They all are part of the network of sites so-called Natura 2000. Similarly, the status of those ISL under a formal protection regulation is good, although some threats persist. This is the case of many lakes in North America and South America: Mono Lake (California, USA), Lake Manitou (Saskatchewan, Canada), Laguna Colorada (Potosi, Bolivia), Mar Chiquita (Cordoba, Argentina), Alchichica (Puebla, Mexico). Advisory or management committees and plans have been established, and some actions are already planned and/or implemented to improve the ecological state of many ISL lakes and their human populations living in the lake shores and watersheds. Nevertheless, the general feeling is that the process of improving their ecological state has been too long for many ISL and that a successful implementation of actions remains to be seen.

4. The likely status of inland salt lakes by 2050 and beyond

A major challenge for improving ISL's ecological state is integrating economic, environmental and social aspects under an efficient governance plan for their management. Such type of plans has been written and implemented for a few lakes but are still missing for many others.

Mono Lake (California, USA) is an excellent example of this approach. Legal protection as a nature reserve was established in 1981, and as scenic area in 1984. A regulatory programme to decrease water diversion from the lake basin was established, and the lake water level was recovering. Also, restoration of degraded streams and lake habitats was performed and, even more, Los Angeles City started a programme to use water more efficiently. The groups promoting the conservation of Mono Lake extended similar programmes for the rational use of water throughout California. However, recovery of the original water level did not proceed as predicted by the hydrologic model, and large areas of the lake shore remain dry. Then, windblown dust causes health problems because of the high concentration of particles in the air, which do not meet the federal PM-10 standard (as in other ISL in North America, Asia and Australia). The challenge now is updating the hydrologic model to achieve a lake water level that does not cause this impact, which will also improve the lake shore habitats and the ecological state of the lake [35].

Aral Sea (Kazajistan/Uzbekistan) still shows a high degradation since the diversion of water from the two major rivers (which also drain areas of other five countries) inflowing water to the lake started promoted by the old URSS before the 1960s. At that time, the natural resources of the Aral Sea were not considered valuable to be protected, neither those in the central Asia desert which was irrigated with the diverted water for agricultural production. After the 1990s, two countries are responsible for the administration of the lake: Kazajistan for the North part and Uzbekistan for the South part. The Kazajistan Government constructed a dam separating both parts of the lake in 2007, and the water level is recovering with improvements in the irrigation system. Further efforts of local groups and international collaboration (both scientific and financial) are contributing to the recovery of fisheries and habitats of the lake. However, corresponding actions have not been implemented in the south part, and degradation progresses with the full desiccation of most of this south part of the lake. Despite the orientations provided by scientists for possible restoration actions to improve the state of Aral Sea [36], the lack of agreement between the two countries managing the lake watershed and others in its watershed is a handicap for the recovery of the Aral Sea. A common plan for these territories could contribute to their sustainable development if the preservation of the Aral Sea values is included.

In 2000 an intergovernmental agreement was signed between two Australian states to maintain the important environmental, social and economic values of Lake Eyre. Later, a collaborative work participated by authorities, environmental groups, local communities and scientists extended the agreement for the sustainable development of the whole lake watershed. All the drivers and pressures listed in **Table 1** are present in Lake Eyre Basin. A very comprehensive plan is applied and renewed every 5 years. It includes the preservation of water flows in the watershed and integrating aboriginal culture and knowledge, promoting a diverse economy adaptable to the effects of climate variability and change [37]. An important point of this plan is that reports of the monitoring and evaluation of the plan actions and lake and basin status will be delivered regularly under the intergovernmental agreement established. These are three examples of the great variety of management practices of inland salt lakes. Those ISL with an integrative management plan and a governance structure for its implementation will have more changes to improve their ecological state.

In any case, climate change impacts and will continue impacting ISL. As they are in arid and semiarid zones, the increasing global temperature enhances evaporation and evapotranspiration. So, a long-term decreasing trend for the water level is observed in many ISL [18, 27, 38]. However, the rainfall pattern is also changing all around and more intense seasonal and interannual patterns (heavier and shorter rainfalls) are being observed in contrast to previous regular longer rain seasons [39]. This climate change feature can negatively affect the biota of a temporary ISL if it dries soon after the rain and there is not enough time for the biota to develop its life cycles, as suggested previously [20]. Also, it can affect positively to temporary ISL with a marked desiccation trend because they will stay flooded for longer periods of time or more frequently within the framework of the long-term water level decreasing trend. However, background climate variation often masks long-term trends in environmental variables, which must be disentangled through robust statistical analysis to attribute lake water level variations to different causes [40].

Revisiting some of the suggestions by Bill Williams and others, **Table 2** shows a few examples of drivers of change for ISL and the responses which would be useful for the improvement of their ecological status.

The first and most important aspect for the management of ISL is considering the management of the lake and its watershed as just one ecosystem. Most of the impacts originated after actions in the lake basin. So, the solution must start by acting in the watershed. Planning and implementing a balanced provision of different ecosystem services in the watershed, covering the human population demands and developing the plan with the information and participation of the population is the first step for the sustainable development of the territory preserving the lake. If this is not incorporated in the management of ISL watersheds, the impacts of land cover and water use changes will continue, and the improvements derived from actions in the lake will not be successful in preserving the natural values of the whole ecosystem.

The perception of the functions and values of ISL is changing all around the world with status of protected sites and conservation and restoration measures already established in many countries of Europe, America and Australia. Even more, many inland salt lakes already have a strategic management plan which includes the ordination of land cover and efficient water use in the watershed. Obviously, this type of plans requires establishing benefits for the local population or facilities to incorporate environmental practices in their current activities. In addition to those cases mentioned above, there are examples of these management practices in China (Lake Qinghai) and Africa (Lake Bogoria). However, one additional requirement for the successful implementation of such type of plans is good communication between stakeholders. This requires mechanisms of participation of local communities to set up the management plan and transparency for the implementation. If all these aspects are not considered in the management plan, the improvement of the ecological state of the lake will not be ensured.

There are examples of ISL with protected status where conservation actions did not progress because of failings to agree such management plan. In Lake Torrens (Australia), an episodic closed lake declared national park, mining operations approved by a local authority were prohibited by the Court of Justice under demand from a local aboriginal community because of damages to cultural sites. This is an example of the need to integrate social aspects while managing ISL. In this case, the conservation of the ISL will benefit from the resolution of the conflict of interests.

Lake Nakuru (Kenya) is well-known for its impressive populations of hippos and flamingoes, which thrive on a simple and very productive trophic web based on microalgae and microcrustaceans. It has been a national park since 1961 but also well-known after the dramatic decrease in flamingoes and fish mortality caused by contamination alerted in the 1970s [41]. Pollution by non-treated wastewater from increasing tourism and the local population, heavy metals from a growing industry and pesticides from expansive agricultural activities [42] jeopardize the conservation of the lake values. Furthermore, siltation caused by deforestation (the basin area covered by natural forest decreased from 47% in 1970 to 8% in 2021) and bad farming practices increased the lake water level, which expanded 60% of the area flooded, impacting the riparian ecology and displacing wildlife, the road network, staff houses, office blocks, electric fences and campgrounds, as well as increasing occurrences of human-wildlife conflict.

The watershed approach – most of the environmental threats are originated in the watershed – is not included yet for the management of many ISL. New delimitations of protected areas incorporating key functional zones of the lake's watershed must be integrated for protection and restoration. For this purpose, integrating the evaluation of ecosystem services [43] is a practical tool to define zones of interest for the ordination of land cover and land uses in spatial planning at a watershed scale. The conservation and management of an ISL can have success if the social, economic and environmental requirements of the whole watershed are satisfied in an integrative way, as socio-ecological system. Additionally, a governance structure must facilitate the participation of stakeholders based on an excellent information system of the lake values and watershed capacities.

The scenarios for the ecological state of inland salt lakes in the future can be outlined by contrasting the effects of climate change (only alternative rainfall is considered as the prediction for temperature is a global increase for the next 30 years, at least) and land cover and water use in the watershed, the two main drivers for the ecological functioning of the lakes (**Table 3**). In general, increased rainfall is not predicted to occur in semiarid and arid zones, but more intense rains and snow events have been observed affecting inland salt lakes [39]. However, warm temperature extremes have also increased, while cold temperature extremes decreased, enhancing evaporation and evapotranspiration, diminishing the water in both the watershed and the lake. Intensification of land cover change and water use diminishing natural habitats will continue the degradation of the lake, whatever the rainfall pattern change. In contrast, a rationale distribution of land cover and water use in the watershed will buffer the impacts of climate change in the lake and, consequently, a relative improvement of the ecological state of the lake will be observed if rainfall decreases or a clear improvement will take place if rainfall increases.

Under these circumstances, the only way to improve the ecological state of an ISL is implementing a sustainable development programme to re-organize watershed land covers and water uses reducing the driver-pressures causing negative impacts in the lake. If this is implemented, restoration of degraded sites in the lake can proceed successfully.

During the present century, new paradigms and approaches have been developed for the management of ecosystems, which are useful for improving the ecological state of ISL. Integrative management plans are useful if implemented at watershed-lake scale with the participation of local communities. For this purpose,

Driver of change for ISL	Case studies	Responses
Climate change	ISL in general	Adaptive management of land and water use in the watershed
Expansion of irrigated agriculture & other water uses in the watershed (urban wastewater discharges)	Aral, Mono, Pyramid, Dead Sea Qinghai, Corangamite, Winnemuca, Owens	Inter-basin collaboration for socio-ecological development. Spatial re-ordination of land covers & land uses. Afforestation (reduced water consumption)
Agriculture irrigation by groundwater pumping in the watershed and inter-basin water exchange	Lakes in Salt Range (Pakistan)	Land cover & Land use reclassification
Clearance of the natural vegetation and other land use changes in the basin	Lake Pyramid	Reforestation/Restoration
Soil erosion in the catchment after deforestation and land consolidation	Lake Gallocanta	Restoration
Urban development in the catchment and lake shores	Great Salt Lake, Lake Nakuru	Urban planning at watershed scale, avoiding taking up the shores of the lake.
Mining	Lake Poopo (water diversion for mining and agriculture & climate change)	Socio-ecological plan after ecosystem services evaluation & adaptation
Exotic species in the lake	Caspian Sea (fish species, invertebrates, introduced)	Removal of invasive species/ Restoration of degraded habitats

Table 2.
Revisiting some of the ideas by bill Williams and others suggesting actions to improve the ecological status of ISL.

		Intensification	Rationalization
Climate change	Decreased rainfall	Severe degradation	Relative improvement
	Increased rainfall	Moderate degradation	Improvement

Table 3.
Likely trends of the ecological state of ISL under different scenarios.

the evaluation of ecosystem services let to define land cover and uses and, finally, a watershed land cover consolidation providing benefits for socio-economic interest and for conservation purposes satisfying the interest of all the stakeholders [44].

A good knowledge of the natural values and processes in the lake and watershed is essential for its management. The essential point is to have a detailed water budget because it lets us know the key inflows and consumption processes of water in the watershed and the lake. Then, management decisions can be taken accordingly to the critical points of the water budget [14]. However, both a short and long-term perspective of the water level changes must be considered as the relative importance of background climate oscillations in ISL remains uncertain and often masks long-term trends in environmental variables but can be accounted for through more comprehensive statistical analyses [40].

Of course, trophic web changes in relation to the water level fluctuations are essential information to know the full spectrum of biodiversity of ISL [45]. This includes the distribution of habitats and communities on the shores of the lake, which also fluctuate as the water level markedly changes [46]. This is fundamental information to anticipate climate change and adopt a strategy to counteract its impacts.

5. Lake Gallocanta as a case study

Lake Gallocanta (22 km^2, watershed 520 km^2, located at 1000 m.a.s.l. at 40° 50' N, 20 11' W in Central Spain) is a closed temporary (maximum water column recorded 250 cm in 1974) salt lake characterized by seasonal and interannual fluctuations of the water level with a return period of 0,3–2,4 years [47]. The lake and 15% of its basin around it were declared protected (Natural Reserve) by the regional Government in 2006. It is a site with natural habitats, vegetation, and animals, protected by the European Union and a Ramsar site.

Lake Gallocanta has two major drivers-pressures in common with many other ISL. Climate change is the major pressure as it has been observed a long-term water level decrease [48], including more frequent desiccated periods during the second half of the twentieth century [49]. Only 1% of the potential water input to the lake is used for irrigation in the watershed [50]. Deforestation for agricultural purposes took place in the mountains surrounding the watershed five decades ago. During the last years, land reparcelling promoted to favor agricultural production has been implemented in most of the municipalities around the lake eliminating hedges and banks, and favoring soil erosion. Also, most of the groundwaters are contaminated by nitrates after years of fertilizing agricultural crops with pig slurry.

The lake maintains a good ecological state, although there is not a current evaluation beyond bird census and reports about the spatial distribution of species and habitats of interest. Recently, an increasing frequency of filamentous algae blooms and oxygen depletion in some parts of the lake bottom linked to heavy rains inflowing suspended solids and nitrates to the lake has been observed. The recently changed rainfall pattern [51] with more intense rainfall events maintains the water fluctuating at a critical intermediate level and salinity. At this stage, a short water level fluctuation significantly changes the trophic structure, which is part of the lake dynamics. The key point is to value all the facies of the lake and their fluctuations in accordance with the natural lake

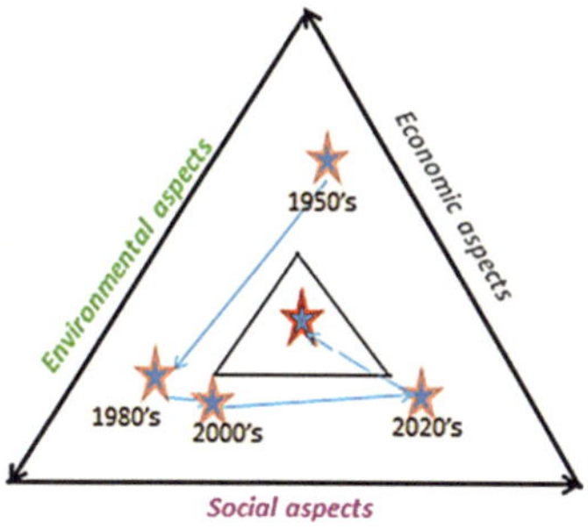

Figure 2.
Changes in the major aspects considered for the management of Lake Gallocanta and its watershed since mid-twentieth century.

dynamics, which in the case of Lake Gallocanta requires desiccation periods following Langbein hydro-ecological model [52] to show all its biodiversity through time.

In the case of Lake Gallocanta, while most of the management attention is devoted to bird populations and bird-watching activities by visitors, a management action plan has not been established yet because of disagreements between farmers and regional Governmental administrators. An erratic management has been taking place with alternative emphasis on different aspects and controversial practices between agricultural and biodiversity measures, although funds provided by the European Union have been given to stimulate agricultural practices in accordance with nature since the mid-twentieth century and for the restoration of natural habitats recently (**Figure 2**).

The history of management of Lake Gallocanta is a paradigm for other lakes. The lack of integration of environmental, social, economic aspects and, overall, lack of governance capacity is a handicap for the conservation of the lake values in the long term. The human population has decreased strongly since the mid-twentieth century, but the economic status of the remaining population did not improve compared to those living in cities. Alternative activities such as tourism or manufacturing did not start up significantly. In addition to specific restoration actions to recover and maintain natural habitats, a management plan integrating the three aspects for the sustainable development of the watershed, including lake conservation, with the participation of local people in the governance programme, is required for the improvement and conservation of the ecological state of the lake.

6. Conclusions

Degradation of ISL progresses all around the world as a consequence of, mostly, impacts originated by land cover and water use changes in the watershed, mining operations and urbanization of lake shores. The status of ISL will improve if management plans are established integrating environmental, social and economic aspects at watershed scale and an efficient governance protocol to implement the management plan. Land cover reparcelling and a rational distribution of water should be part of the management plan. Adaptive management practices in the watershed are required to buffer climate change impacts. Continuous control of the lake's physical, chemical and biological conditions will provide the required information for the evaluation of their status and innovation of the management plan.

Author details

Francisco A. Comín
Pyrenean Institute of Ecology-CSIC, Jaca/Zaragoza, Spain

*Address all correspondence to: comin@ipe.csic.es

References

[1] Williams WD. Problems in the management of inland saline lakes. Verhandlungen des Internationalen Verein Limnologie. 1981;**21**:688-692

[2] Zadereev E. Salt Lake ecology and environment. In: Zheng M, Deng T, Oren A, editors. Introduction to Salt Lake Sciences. Beijing: Science Press; 2018. pp. 172-179

[3] Young RT. The life of Devils Lake. North Dakota: North Dakota Biological Station. Pub; 1924. p. 113

[4] Jenkin PM. Biology of lakes in Kenya. Nature 1932; 124 (Num. 3128):574-574.

[5] Arnold R, Johnson HR. Sodium sulphate in soda Lake, Carrizo plain, San Luis Obispo country, California. U.S. Geological Survey. 1909;**380**:369-371

[6] Teeple JE. The Industrial Development of Searles Lake Brines with Equilibrium Data. American Chemical Society Monograph Series. New York: Chemical Catalog Co. Inc.; 1929. p. 182

[7] Hammer UT. Saline Ecosystems of the World. Dordrecht: Dr. W. Junk Publishers/Kluwer Academic Publishers Group; 1986. p. 616

[8] Jellison R, Zadereev YS, DasSarma PA, Melack JM, Rosen MR, Degermendzhy AG, et al. Conservation and management challenges of Saline Lakes: A review of five experience briefs. Lake Basin Management Initiative, Thematic Paper. 2004. p. 28. Available from: https://www.researchgate.net/publication/228885117

[9] Timms BV. Salt lakes in Australia: Present problems and prognosis for the future. Hydrobiologia. 2005;**552**:1-15

[10] Williams WD. Inland salt lakes: An introduction. Hydrobiologia. 1981;**5**:1-14

[11] Williams WD. Limnological imbalances: An antipodean viewpoint. Freshwater Biology. 1988;**20**(3):407-420

[12] Williams WD. What future salt lakes? Environment. 1996;**38**(9):12-20

[13] Eugster HP, Hardie LA. Saline Lakes. In: Lerman A, editor. Lakes. New York, NY: Springer; 1978. pp. 237-293

[14] Wurtsbaugh WA, Miller C, Null SE, DeRose RJ, Eilcock P, Hahnenberger M, et al. Decline of the world's saline lakes. Nature Geoscience. 2017;**10**:816

[15] Jellison R, Williams WD, Timms B, Alcocer J, Aladin AV. Salt lakes: Values, threats and future. Aquatic ecosystems: Threats and future. In: Polunin NVC, editor. Aquatic Ecosystems: Trends and Global Prospects. Cambridge: Cambridge University Press; 2008. pp. 94-110. DOI: 10.1017/CBO9780511751790.010

[16] Heydari N, Jabbari H. Worldwide environmental threats to salt lakes. International Journal of Design & Nature and Ecodynamics. 2012;**7**(3):292-299

[17] Comin FA, Williams WD. Parched continents: Our common future? In: Margalef R, editor. Limnology Now: A Paradigm of Planetary Problems. New York: Elsevier Science B.V; 1994. pp. 473-527

[18] Mason IM, Guzkowsk MAJ, Rapley CG. The response of lake levels and areas to climatic change. Climate Change. 1994;**27**:161-197

[19] Williams WD. The largest, highest and lowest lakes of the world: Saline

lakes. Verhandlungen des Internationalen Verein Limnologie. 1996;**26**:61-79

[20] Williams WD. Environmental threats to salt lakes and the likely status of inland saline ecosystems in 2025. Environmental Conservation. 2002;**29**(2):154-167

[21] Van Eck N, Waltman L. Software survey: VOSviewer, a computer program for bibliometric analysis. Scientometrics. 2010;**84**(2):523-538

[22] Smeets E, Weterings R. Environmental indicators: Typology and overview. Copenhagen: European Environment Agency; 1999. p. 19

[23] Hussain SA, Han FQ, Han W, Rodríguez A, Han JL, Han J, et al. Climate change impact on the evolution of the Saline Lakes of the Soan Sakaser Valley (central salt range; Pakistan): Evidences from hydrochemistry and water (δD, δ 18O) and chlorine (δ 37Cl) stable isotopes. Water. 2019;**11**:912. DOI: 10.3390/w11050912

[24] List of Drying Lakes. 2023. Available from: https://en.wikipedia.org/wiki/ [Accessed: February 1, 2023]

[25] Street FA. The relative importance of climate and local hydrogeological factors in influencing Lake level fluctuations. Paleoecology In Africa. 1980;**12**:137-158

[26] Evans T. The effects of changes in the world hydrological cycle on availability of water resources. In: Bazzaz F, Sombfroek W, editors. Global Climate Change and Agricultural Production. Direct and Indirect Effects of Changing Hydrological, Pedological and Plant Physiological Processes. Chichester: FAO & John Wiley & Sons; 1996

[27] Zadereev E, Lipka O, Karimov B, Krylenko M, Elias V, Pinto IS, et al. Overview of past, current, and future ecosystem and biodiversity trends of inland saline lakes of Europe and Central Asia. Inland Waters. 2020;**10**:438-452

[28] Woolway RI, Kraemer BM, Lenters JD, Merchant CJ, O'Reilly CM, Sharma S. Global lake responses to climate change. Nature Reviews Earth Environment. 2020;**1**:388-403

[29] Golubev GN. Caspian and Aral Sea: Two different paths of environmental degradation. Verhandlungen des Internationalen Verein Limnologie. 1996;**26**(159):166

[30] Timmer KL, Wallace L. Coordinated Watershed Management Strategy for the Middle Truckee River. 2004. Truckee:Truckee River Watershed Council; 2004. p. 236. [www.truckeeriverwc.org]

[31] Wurtsbaugh WA, Sima S. Contrasting management and fates of two Sister Lakes: Great salt Lake (USA) and Lake Urmia (Iran). Water. 2022;**14**(19):3005. DOI: 10.3390/w14193005

[32] Johnston J, Razafy M, Lugo H, Olmedo L, Shohreh S, Farzan F. The disappearing Salton Sea: A critical reflection on the emerging environmental threat of disappearing saline lakes and potential impacts on children's health. Science of the Total Environment. 2019;**663**:804-817

[33] Chang B, He K, Li R, Sheng Z, Wang H. Linkage of climatic factors and human activities with water level fluctuations in Qinghai Lake in the Northeastern Tibetan plateau, China. Water. 2017;**9**(552):1-13. DOI: 10.3390/w9070552

[34] Marconi P, Arengo F, Clark A. The arid Andean plateau waterscapes and the lithium triangle: Flamingos as flagships

for conservation of high-altitude wetlands under pressure from mining development. Wetlands Ecology and Management. 2022;**30**:827-852

[35] History of the Mono Lake Committee. Available from: https://www.monolake.org/today/raising-mono-lake-requires-a-new-plan [Accessed: February 2, 2023]

[36] Dumetz J, Aladin N. Saving the Aral Sea: A cross-cultural communication success. Academia Letters. 2021;**360**:1-4. DOI: 10.20935/AL360

[37] DCCEEW. Draft Lake Eyre Basin Strategic Plan. Canberra: Department of Climate Change, Energy, the Environment and Water; 2022. p. 17 Available from: https://dcceew.gov.au/publications

[38] Meng Q. Climate change and extreme weather drive the declines of Saline Lakes: A showcase of the great salt Lake. Climate. 2019;7(19):1-12. DOI: 10.3390/cli7020019

[39] Zhang W, Furtado K, Wu P, Zhou T, Chadwick R, Marzin C, et al. Increasing precipitation variability on daily-to-multiyear time scales in a warmer world. Science Advances. 2021;7:eabf8021

[40] Kraemer BM, Seimon A, Adrian R, McIntyre PB. Worldwide lake level trends and responses to background climate variation. Hydrology and Earth System Sciences. 2020;**24**:2593-2608

[41] Koeman JH, Pennings JH, de Goeij JJM, Tjioe PS, Olindo PM, Hopcraft J. A preliminary survey of the possible contamination of Lake Nakuru in Kenya with some metals and chlorinated hydrocarbon pesticides. Journal of Applied Ecology. 1972;**9**(2):411-416

[42] Tenai BC, Mbaria JM, Muchemi GM, Jansen R, Kotze A, Naidoo V, et al. Assessment of heavy metals concentration in water, soil sediment and biological tissues of the lesser flamingos in four eastern rift valley lakes. African Journal of Environmental Science and Technology. 2016;**10**(6):162-166

[43] Millennium Ecosystem Assessment. Ecosystems and Human Well-Being: A Framework for Assessment. Washington DC: Island Press; 2003. p. 245

[44] Comín FA, Miranda B, Sorando R, Felipe-Lucia MR, Jiménez JJ, Navarro E. Prioritizing sites for ecological restoration based on ecosystem services. Journal of Applied Ecology. 2018;**55**(3):1155-1163

[45] Barrett KL, Belosky GE. Invertebrates and phytoplankton of great salt Lake: Is salinity the driving factor? In: Baxter BK, Butler JK, editors. Great Salt Lake Biology, a Terminal Lake in a Time of Change. Cham: Springer Nature Switzerland; 2020. pp. 145-173. DOI: 10.1007/978-3-030-40352-2

[46] De Deckker P. Biological and sedimentary facies of Australian salt lakes. Palaeogeography, Palaeoclimatology, Palaeoecology. 1988;**62**:237-270

[47] Comin FA, Julia R, Comin P. Fluctuations, the key aspect for the ecological interpretation of saline lake ecosystems. Oecologia Aquatica. 1991;**10**:127-135

[48] Pérez ME, García MP. Fluctuaciones hídricas de la laguna de Gallocanta analizadas a partir de teledetección. Ería. 2015;**96**:97-106

[49] Rodo X, Baert E, Comin FA. Variations in seasonal rainfall in

DOI: http://dx.doi.org/10.5772/intechopen.111447

Southern Europe during the present century: relationships with the North Atlantic Oscillation and the El Niño-Southern Oscillation. Climate Dynamics. 1997;**13**:275-284

[50] García MA, San Román J, Blasco O, Coloma P. Hydrogeology of the Gallocanta Lake and environmental implications. Memorias de la Real Sociedad Española de Historia Natural. 2009;**7**(2ª):77-101

[51] Kuhn NJ, Baumhauer R, Schütt B. Managing the impact of climate change on the hydrology of the Gallocanta Basin, NE Spain. Journal of Environmental Management. 2011;**92**:275-283

[52] Langbein WB. Salinity and hydrology of closed lakes. In: Geological Survey Prof. Paper 412. Washington: U.S. Govt. Print. Off; 1961. p. 20

Chapter 4

Economic Impacts of the Establishment of Alternative Water Retention Habitats on the Agricultural Holdings

Matjaž Glavan

Abstract

The municipal spatial plan of the city of Ljubljana determined the location for the expansion of the Ljubljana Regional Waste Management Centre. The environmental condition for the expansion is the establishment of alternative water retention habitats (267.1 ha) on the northern edge of the Ljubljana Marsh. The study aimed to analyze possible mitigation measures for agriculture due to the envisaged changes in land use. The plan envisaged a change in the use of conventional agricultural land, overgrowth vegetation and forest for extensive grassland (172 ha), forest (86 ha), and water surfaces (8.9 ha). Results indicate that the income from subsidy payments will be higher due to establishing additional wetland meadows (28 ha) in the currently overgrown areas. The intervention will decrease the value of the crop produce (−61 to −71%) and thus the farm revenue (−34 to −43%). However, variable costs are lower due to the expected extensive land use. A larger area of protected habitats on agricultural land returns lower variable costs (−60 to −69), a positive balance of gross margin for the total area (+5 to +15%), and a lower gross margin per hectare of agricultural land (−4 to −12%), and thus the income of most agricultural holdings.

Keywords: alternative habitat, water retention, agriculture, income, wetland, economic calculation

1. Introduction

Agriculture has an important impact on biodiversity; both are interconnected through ecological functions and ecosystem services, such as soil structure, nutrient content, pollination, regulation of pests and diseases, water retention, and semi-natural habitats species, depending on Ref. [1]. With appropriate policies, it is possible to maintain and even enhance biodiversity while agriculture adjusts cultivation practices. However, the cost-effectiveness of different policies needs to be tested before implementation [1].

In most countries, national laws on nature conservation specify mitigation and compensatory measures to mitigate the obstruction of nature or its consequences or to compensate for the intended or caused degradation of nature. Among the possible forms of these measures is establishing an alternative nature area with the same nature conservation characteristics as the area on which the intervention had a significant impact [2].

The effects of the establishment of alternative water retention areas and extensive wet grassland habitats can affect agriculture in several ways. The effects can be divided into direct and indirect, depending on the consequences. The direct adverse effects on agricultural land include a change in soil properties and, thus, a reduction in the production potential of agricultural land. This is due to the flooding of land and the habitat requirements of plant and animal organisms, which in a given case, require conditions similar to wetlands or wet meadows. The direct effects also include the loss of agricultural land due to the establishment of surface water bodies with permanent or intermittent standing water for habitat needs and surfaces for flood water storage. Indirect adverse effects include impacts that originate from changes in soil properties (soil temperature and soil water content). As a result, problems arise in soil cultivation (access to land), loss of yield and income, and land value.

On surfaces with a high-water table level, such as wet meadows, the water-saturated soil area is closer to the soil surface. All soil horizons are entirely saturated with water during most of the year. It creates special conditions in the soil that affect the growth of agricultural plants typical of the climate of the study area. Most agricultural plants in waterlogged soils lack oxygen for root growth. It should be emphasized that higher groundwater levels or deepening the terrain to retain flood water on agricultural land represents a reduction in the production potential of the agricultural land for crop production, regardless of whether market or nonmarket crop production takes place on the agricultural land. In order to achieve the suitability of the wet meadow habitats for the life of classification bird species, a constant high-water level in the area of wet meadows establishment is necessary. Lower-water levels are foreseen only during grass harvesting. By regulating the water regime, we are not establishing a natural system but an artificially made system that needs constant maintenance.

The location for the expansion of the Regional Waste Management Centre (RCERO) was determined by the municipal spatial plan of the city of Ljubljana. The spatial plan determines that one of the conditions for expanding the RCERO is establishing alternative areas for retaining flood water and habitats of extensive wet meadows on the northern edge of the Ljubljana Marsh.

The objectives of this study are (i) an analysis of the economic impact of establishing alternative water retention habitats on the income from agricultural activity and (ii) the analysis of possible mitigation measures for agriculture due to the establishment of alternative habitats. The study presents a spatial analysis of the existing and alternative economic situation analyzing the change in revenues, costs, and gross margins per hectare of agricultural land due to the establishment of alternative water retention habitats for the three proposed land use variants [3, 4].

2. Methodology and materials

2.1 Study area

The area proposed for implementing measures to establish alternative or replacement water retention habitats to ensure a favorable condition of the qualification species of Eurasian woodcock (*Scolopax rusticola*) was established to mitigate the expansion of the RCERO. The area is located in Slovenia in the municipality of

Ljubljana in the northern part of the Ljubljana Marsh, south of the A1/A2 highway, east of the regional center for waste management, west of Rakova Jelša settlement and north of the Ljubljanica river (**Figure 1a**). The area of 267.1 ha is flat, with an altitude between 277 and 278 m.

The climate in the area is classified as moderate continental of Central Slovenia, or the so-called sub-alpine climate [5]. Ljubljana lies in an extensive basin surrounded by the pre-alpine and Karst geology. The temperature inversion is characteristic of the area. The marsh is a source of cool air in the summer, as temperatures are, on average, four to five degrees lower than in the city [6]. Average annual weather parameters in Ljubljana for the period 1971–2000 are (i) temperature of 10.9°C, (ii) 1974 hours of sunshine, and (iii) 1362 mm of precipitation.

According to the soil map, the more significant part of the soil in the area is classified as low peat marshes, which are shallow to medium deep humified (40%)

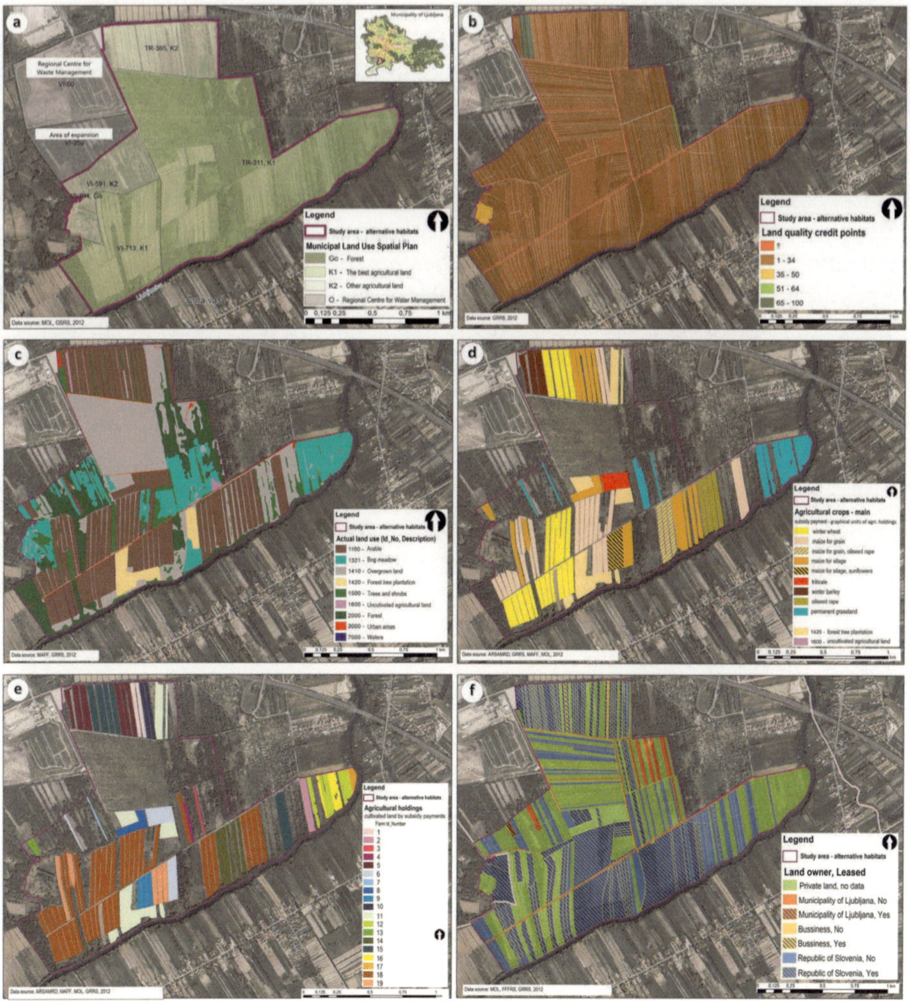

Figure 1.
Research area (a) land use by municipal spatial plan, (b) land quality credit points, (c) actual land use, (d) land use graphical units of agricultural holdings (GERK) with main crops in 2012, (e) land cultivated by agricultural holdings, and (f) ownership and lease of the land.

and mineral-organic clay loams of calcareous origin (38%). Other soils are strongly expressed gleysol on the organic subsoil (15%) and eutric mineral, medium to strongly expressed hypogleys (5%) [7]. The largest share of the area is represented by the land quality credit points class of 1–34 points with 255.31 ha (95.6%). Of this, 49.1% or 131.21 ha is agricultural land with a rating of 13 points (**Figure 1b**).

From the total area of 267 ha, in terms of actual land use, arable land presents 34% (92 ha), overgrown agricultural land 26% (70 ha), marshy wet meadows 14% (38 ha), forest 14% (37 ha), trees and shrubs 7% (18 ha), and forest tree plantations 3% (8 ha). The rest is uncultivated agricultural, urbanized land, and water surfaces (**Figure 1c**). There are 72 land-use graphical units of agricultural holdings (GERK) (109.85 ha), presenting 41% of the total study area (**Figure 1d**). Among all GERKs, 52 have arable land use (85.09 ha), 30 marsh/bog meadow land use (10.89 ha), and 7 forest tree plantation land use (10.53 ha).

Data on subsidy payments for agriculture shows that 98.27 ha of agricultural land was claimed in the study area (greening is not included). The agricultural crop code permanent meadow (204) is attributed to 12 GERKs (13 ha) (**Figure 1d**). Agricultural crop codes attributed to fields (001, 005, 006, 007, 009, 012, 014) in 2011 and 2012 were used on 52 GERKs (85.18 ha). In an average year, the highest share of land was planted by winter wheat, followed by silage maize, grain maize, oilseed rape, and other crops (**Table 1**). On average, other crops occupied less than 5% of the surface. The most common rotations in the area are two-year rotations (maize/wheat or barley or triticale), three-year rotations (maize/maize/wheat or triticale or barley), and four-year rotations (maize/oilseed rape/maize/wheat or triticale or barley).

Agricultural land was cultivated by 19 holders of agricultural activity registered in the register of agricultural holdings, which are included in the system of subsidy payments (**Figure 1e**). All agricultural holdings cultivated 719 ha of agricultural land of which 109.55 ha, or 15.24%, lie within the study area (**Figure 2**). Two agricultural holdings owned more than 50% of the agricultural land within the study area.

Crop		**Area (ha)**			
		Year 2011	**Year 2012**	**Average**	
Id	Name		Ha		%
001	winter wheat	17.98	32.31	25.15	22.52
005	maize—grain	19.06	22.92	20.99	18.80
006	maize—silage	19.99	23.06	21.53	19.28
007	triticale	2.31	2.60	2.46	2.20
009	winter barley	3.96	4.46	4.21	3.77
012	sunflower	2.12	0.99	1.56	1.39
014	oilseed rape	11.67	25.63	18.65	16.70
204	permanent grassland	12.54	13.70	13.12	11.75
206	clover-grass mixture	5.96	—	5.96	5.34
207	clover	2.05	—	2.05	1.84
	Sum	97.64	97.64	125.67	111.66

Table 1.
Areas (ha) of crops on fields for the research area, including greening, based on subsidy payments in 2011 and 2012.

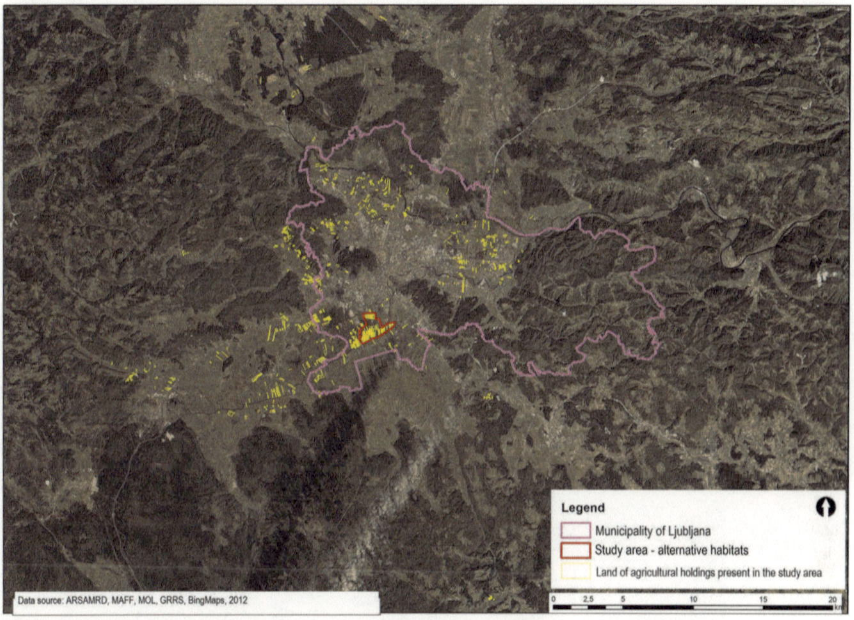

Figure 2.
All agricultural land cultivated by agricultural holdings present in the study area of alternative water retention habitats.

Agri-environmental measures (AEM) from the Rural Development Program (RDP) of the Republic of Slovenia (2007–2013) were implemented as the sub-measures of sustainable livestock production (SLP) on 7.5 ha, crop rotation (ROT) on 45.2 ha, and greening of arable land (GAL) on 26.38 ha. In addition, compensatory payments for less favored areas with limited opportunities for agricultural activity were applied to 88.3 ha. There are 374 land parcels in the area, most of which are owned by the Republic of Slovenia (146.94 ha, 55%) (**Figure 1f**).

There are no official drainage or irrigation systems in the area (**Figure 3**). The area is otherwise characterized by open drainage ditches, which drain excess soil and groundwater from the area and directly enable the cultivation of agricultural land. The area has three main drainage ditches: Curnovec, Lahov Graben, and Kansov Graben. The southern part of the area, which is agriculturally intensive, is exposed to frequent floods, the central part of rare floods, and the northern part to catastrophic floods. The high flood risk area covers 98 ha or 38% of the study area (south), the medium flood risk area covers 108 ha or 42% of the area (central), and the low flood risk area covers 44 ha or 17% of the area (north). The entire study area is classified as a NATURA 2000 area.

2.2 Data source

Both spatial and tabular data from various public sources are included in the analysis (**Table 2**). We edited the data using ESRI® ArcGIS 10.0 software (ArcMap, ArcInfo, ArcCatalog) and MS Excel. Using ArcGIS, we combined the tabular data with spatial layers and obtained additional information about the study area. Data are displayed spatially and tabularly with appropriate written explanations.

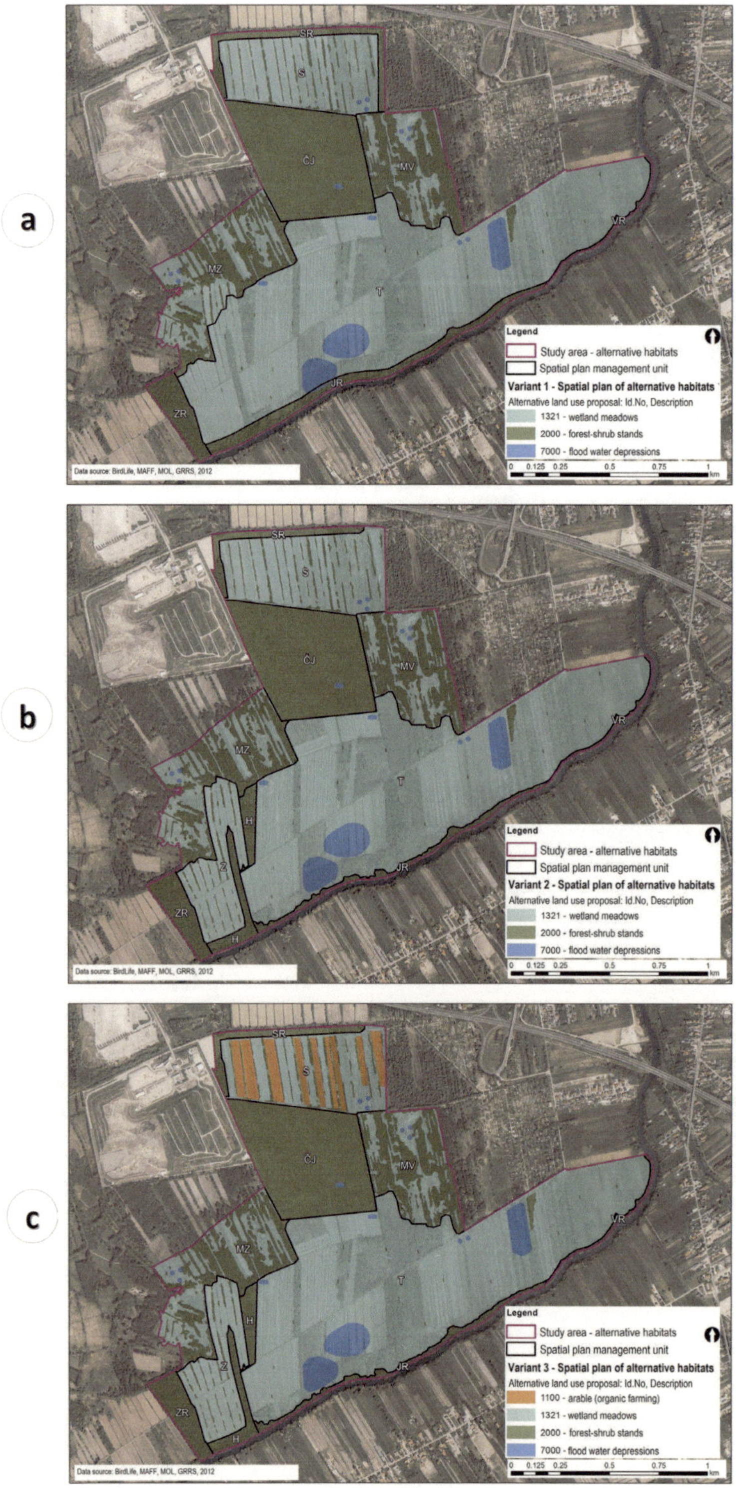

Figure 3.
Land use of three variant solutions for the establishment of alternative water retention habitats in the study area with the presented spatial management units.

Data	Type	Source
Municipal Spatial Plan of the City of Ljubljana (OPN); Flood risk; Water protection areas; Natura 2000; Ecologically important areas; Natural heritage—Landscape Park of the Ljubljansko barje; Cultural Heritage.	Spatial layer Tabular data	City Municipality of Ljubljana (MOL), Spatial planning department (2012); Ministry of the Environment and Spatial Planning (MESP) (2012).
Pedocartographic units; Digital soil number; Actual land use; Land use graphical units of agricultural holdings; Drainage and irrigation systems.	Spatial layer Tabular data	Ministry of Agriculture, Forestry and Food (MAFF) (2012) http://rkg.gov.si/GERK/
Land quality credit points State land lease contracts	Tabular data	Geodetic Survey of the Republic of Slovenia (GSRS) (2012) http://www.gu.gov.si/ Farmland forest Fund (FFFRS)
Subsidy payments Type of agricultural culture Agri-environmental measures payments Agricultural holdings	Tabular data	Agency of the Republic of Slovenia for Agricultural Markets and Rural Development (ARSAMRD) (2012) http://www.arsktrp.gov.si/si/

Table 2.
Data sources included in the analysis.

The methodology includes calculating the potential change in gross margin (gross margin = revenue – variable costs) due to changes in production and thus revenue and costs from agricultural activities for the area of permanent agricultural land occupation (eqs. 1 and 2). Permanent occupation means that agricultural land is managed by state policy-determined regulations under a special regime to establish alternative or replacement habitats and water surfaces to provide an alternative volume for retaining flood water. In the assessment of economic effects, we included: (a) the spatial location of the agricultural land in the study area; (b) agricultural land use, type of agricultural culture, and crop rotation; (c) three variant solutions for establishing alternative wetland habitats and replacement volume for flood waters; (d) involvement in agri-environmental measures (AEM) of the EU Common Agricultural Policy Rural Development Program.

An assessment of the change in economic parameters due to the permanent occupation of agricultural land of various types was made. Conventionally managed arable fields, marsh meadows, and forest tree plantations (**Figure 1c**) are to be replaced to a lesser or greater extent by types of extensive-organic agricultural land use (organic arable fields, marsh meadows), forests, trees and shrubs, and water surfaces. Changes in land use mean a change in the type of plants, the quantity of the crop yield, and crop quality. The potential change in gross margin for the average growing season and the average rotation was assessed, considering the Catalog of calculations for selected crops [8]. The catalog is also used by the Slovenian Chamber of Agriculture and Forestry and the Ministry of Agriculture, Forestry and Food (**Tables 3–5**).

From the calculations for individual crops, we prepared two average calculations for arable areas, which include the basic average values of revenue and variable costs

ARABLE FIELD, three-year rotation (K3), conventional farming winter wheat/grain maize/silage maize	**Calculation (EUR/ha/year)**
Value of the produce (EUR/ha)	1091.7
Subsidy payment (EUR/ha)	332.0
Revenue (EUR/ha)	1423.6
Variable Costs (EUR/ha)	1040.3
Gross margin (EUR/ha)	383.3
ARABLE, three-year rotation—organic—OA winter wheat/buckwheat/ grain maize	
Value of the produce (EUR/ha)	1225.0
Subsidy payment (EUR/ha)	630.0
Revenue (EUR/ha)	1855.0
Variable Costs (EUR/ha)	1031.3
Gross margin (EUR/ha)	823.7

Table 3.
Calculation of variable costs and gross margin for arable crop rotations.

Wood of non-forest plantation-P 15 years life span of the plantation. Chips, chopping every 5 years, 7 t dry matter/ha/year	**Produce price (€/t)**	**Calculation (average yield)**
Yield (m^3/ha)		7
Value of the produce (€/ha)	38	266
Subsidy payment (EUR/ha)		—
Revenue (EUR/ha)	38	266
Variable Costs (EUR/ha)		140
Gross margin (EUR/ha)	38	126

Table 4.
Calculation of variable costs and gross margin for the wood of non-forest plantation.

for calculating gross margin (**Table 3**). The basis for preparing these calculations was based on an average three-year rotation (wheat/grain maize/silage maize). For the alternative habitat variant 3, which envisages the preservation of 10 ha of organically managed arable fields, we used an organic three-year rotation (wheat/buckwheat/grain maize).

For crops sold on a larger scale, the average produce prices detected on the market in the year of publication of the Catalog of Calculations are considered and, in some instances, do not reflect the actual situation [8]. Furthermore, prices on the market fluctuate annually and monthly depending on supply and demand, which means that the products can be sold at a higher or lower price than considered in the calculation.

The economic calculation for agricultural holdings is based on the subsidy payments applications at the Agency for Agricultural Markets and Rural Development. In order to compare the differences between the existing land use situation (ExU) and the three extensive variants of the alternative replacement habitats (V1, V2, V3) and

HAY, unfertlizied—conventional—M1 dried on the soil, baled, 2 cuts*; 85% final dry matter	**Produce price (€/t)**	**Calculation (average yield)**
Yield green (t/ha)		30.0
Yield hay (t/ha)		6.3
Value of the produce (€/ha)	87.34	550.2
Subsidy payment—REG (€/ha)		109.0
Subsidy payment—BH (€/ha)		121.4
Revenue (€/ha)		780.6
Variable costs (€/ha)		240.0
Gross margin (€/ha)		540.6
HAY, unfertlizied—organic—M3 dried on the soil, baled, 2 cuts*; 85% final dry matter; the first cut after July 1st—1/3 of the conventional one in terms of quality		
Yield hay (t/ha)		6.3
Value of the produce (€/ha)	52.6	331.5
Subsidy payment—REG and OA (€/ha)		336.0
Revenue (€/ha)		667.5
Variable costs (€/ha)		240.0
Gross margin (€/ha)		427.5
HAY, unfertlizied—organic—M5 dried on the soil, baled, 1–2 cuts; 85% final dry matter; first year, one cut after August 1st, suitable for bedding; the second year, two cuts—the first cut after July 1st, which is 1/3 of the conventional one in terms of quality		
Yield hay (t/ha)		5.1
Value of the produce (€/ha)	26.3	165.8
Subsidy payment—REG and OA (€/ha)		336.0
Revenue (€/ha)		501.8
Variable costs (€/ha)		178.8
Gross margin (€/ha)		323.0

**ratio in the hay crop ➔ first cut: second cut = 60:40.*
REG—subsidy payment for grassland; OA—agri-environmental measures of organic agriculture; BH—agri-environmental measures of preservation of butterfly grassland habitats, grass cut before July 1st and after August 20th.

Table 5.
Calculation of variable costs and gross margin for an unfertilized marsh meadow, where dried hay is grown for fodder, produced as bales in conventional and organic agriculture.

their economic effects on agriculture, in the case of arable land, meadow land, and forest tree plantations, we used data from the calculation tables (**Tables 3–5**) on the average annual produce yield (PY) and crop price (CP), average annual subsidy payments (NP), average annual revenue at a given price (R), average annual variable costs (VC), and average annual gross margin at a given price (GM) (**Tables 3–5**). The calculation does not include subsidy payments for less favored areas for agricultural activity (OMD), which greatly vary between agricultural holdings and differ depending on the average quality of land and its geographic position.

Equations and terms used in the calculation of the economic effects of the arrangement of alternative habitats on agricultural land:

$$GM = R - VC. \tag{1}$$

$$R = (PY \times CP) + SP \tag{2}$$

where

Gross margin (GM) = revenue [€/ha] – cost [€/ha].

Revenue (R) = (crop price [€/ha] × produce [t/ha]) + subsidy payments [€/ha] = [€/ha].

Variable cost (VC) = seeds, fertilizers, pesticides, machine hours, work hours [€/ha].

Produce yield (PY) = average harvested crop yield [t/ha].

Crop price (CP) = price of produce [€/ha].

Subsidy payments (SP) = EU Common Agricultural Policy Rural Development Program (2007–2013) Republic of Slovenia [€/ha].

2.3 Variant proposals for the establishment of alternative habitats

We designed three types of economic effects simulations described in the report entitled variant proposals for arranging alternative habitats due to the expansion of the Ljubljana landfill [3, 4]. The report presents three proposals for placing alternative habitats and replacing volumes for flood waters (**Figure 3**). Variants 1 (V1) and 2 (V2) propose that currently as marsh meadows (81.9 ha), forest (4.8 ha), and water surfaces (5.6 ha), with a land use ratio of 32.3:64.4:3.3. Variant 3 (V3) propose arable land (92.3 ha) to be managed organic fields (9.7 ha), marsh meadows (74.8 ha), forest (4.3 ha), and water surfaces (3.4 ha), with land use ratio of 32.3:60.3:3.3:4.1.

The proposed variants V1 and V2 fully follow the provisions of the Municipal Spatial Plan. In both variants, forest-shrub and meadow areas were arranged in a mosaic in the prescribed ratio of 30%:70%. The only differences are in the location of the forest-shrub vegetation on the northern edge of the area and the southern strip of existing riparian trees along the Ljubljanica River. The proposed variant 3 (V3) differs from V2 in approximately 11 ha of organically managed arable land for the production of low-growing agricultural crops (close-grown cereals, root crops). It prohibits the cultivation of maize and other tall grains. The use of plant protection products is restricted. It also envisages a second location for the floodwater depression in the eastern part, placing it on overgrown land. The most significant impact on agriculture is expected in areas where the proposal envisages forest-shrub stands and floodwater depressions. In those, agricultural activity will be completely disabled, which means that these agricultural lands will be permanently taken away or permanently occupied with alternative habitats, serving as floodwater retention areas during floods and as water surfaces for birds in normal conditions. These areas will be redesigned and deepened. To provide optimal conditions for the life of Natura 2000 classified plant and animal species, water in floodwater depressions will be present for most of the year, or the groundwater level will be so high that it will not allow agricultural activity.

In the area of management units (V1: MZ, MV; V2-V3: MZ, MV, H), where the establishment of mosaic land use pattern is envisaged, forest-shrub stands and wet marsh meadows are planned, which are not fertilized and are mowed maximal twice a year (**Figure 3**). It is also envisaged that the management of the drain ditches network and overgrowth of the ditches that separate the plots will be discontinued. Therefore, agricultural activity will be limited in the wet marsh meadow management units (V1: T, S; V2-V3: T, S, Z) (**Figure 3**).

In the unit T area, all arable land will be converted to unfertilized alternately cut meadows, and all tree and shrub growth will be removed except for a few clusters of shrub vegetation. It is envisaged that the ditches will be preserved, but they will be equipped with sluice gates to maintain an appropriate groundwater level. In the eastern part of unit T, sheep grazing is expected to be preserved, but only if there are no adverse impacts on the classification species of Natura 2000.

All fields in management units S and Z will be turned into unfertilized, twice-a-year cut marsh meadows. It is envisaged that the parcel borders will be overgrown, and 10 clusters of shrubs and tree vegetation will be planted. It is also planned to discontinue the management of the drainage ditches to increase the soil moisture of the land. However, it foresees the renovation of the drainage ditches and the installation of sluice gates if mechanical grass cutting is impossible.

Agricultural activity will not be possible in units with forest cover and shading vegetation (V1: ZR, SR, JR, VR; V2-V3: ZR, SR, JR, VR, H) (**Figure 3**). The forest will be excluded from management and left to natural development. Most forest and shrub areas already exist, so no significant impact on agricultural activity is expected. The only exception is the northern edge of management unit S, where a new riparian strip of vegetation is established on agricultural land.

Variant 3 in management unit S proposed (**Figure 3**) organically managed crop production, no pesticides, low-growing cereals with o maize, and the preservation and maintenance of drainage ditches. According to BirdLife Slovenia, the complete ban on the cultivation of maize is related to the shape of the maize stand, which is not optimal habitat or has a negative impact on many Natura 2000 classification bird species. Based on our discussion with bird experts, we have included maize in the organic rotation (11 ha of fields in the area) due to its beneficial effect on production economics. However, with one condition, only organically produced local maize varieties can be grown in an alternating rotation with winter wheat and buckwheat.

From the description of the proposed management in the study area, it is clear that any agricultural activity will be limited and, in some places, even impossible compared to the existing practice after establishing alternative habitats and flood water retention areas. The proposed management restricts not only conventional but also organic cultivation. Production on arable fields is mainly prohibited, as is the fertilization of grassland areas. If a sluice gate system is established in the area to maintain a higher water table level and the water table is not higher than 20 cm below the ground level, this could favor grass production. However, it should be noted that grass from single-cut, late-season cut meadows is not suitable for animal consumption in intensive agricultural production. The fodder is only suitable for bedding or as supplementary fodder for horses.

3. Results and discussion

3.1 Existing situation

The existing land use situation results show that the average annual revenue (value of the crop and subsidy payments without agri-environmental measures) for the three-year rotation varies around 164,425 EUR per year for the entire study area (**Table 6**). The produced crop value is estimated at over 125,000 EUR per year. Disregarding payments for less-favored areas and agri-environmental measures (AEM), the maximum possible annual amount of subsidy payments per area is estimated at over 39,000 EUR per year. Considering the AEM payments, a minimum of around 9000 EUR per year can be added to the revenue. A realistic average estimate of AEM payments is difficult, as it changes from year to year and depends on the agricultural policy and the voluntary willingness of farmers to join the AEM scheme. Variable costs average around 107,000 EUR per year. Thus, the average gross margin for agricultural land in the study area is estimated at 57,418 EUR per year. If we add the AEM payment, the estimated gross margin increases to a minimum of 66,684 EUR per year.

As already mentioned, the calculation did not include payments for less-favored areas and AEM payments, such as the preservation of the crop rotation (ROT) (91.84 EUR per ha annually), the greening of arable areas (GAL) (172.20 EUR per ha annually), and sustainable livestock production (SLP) (84.46 EUR per ha annually). If the entire study area fields were included in AEM, revenue would increase substantially (ROT by 8477 EUR annually; GAL by 15,894 EUR annually) and thus also the total coverage. However, receiving payments is conditioned with a required five-year crop rotation, which not all farmers can agree to due to the farming type specificity and cultivation technology. In the calculation, the AEM preservation of butterfly grassland habitats (BH) is considered, stipulating that grass cutting must not be done between July 1st and August 20th, enabling two quality grass cuts. In the study area, all agricultural holdings practice conventional production. Therefore, none of them applied for AEM payments for organic agriculture (OA).

3.2 Economic impact of alternative habitats on agriculture

The results of the analysis of alternative habitat variants 1, 2, and 3 show that total revenues would decrease by 43% (V1), 41% (V2), and 34% (V3), respectively, mainly due to a lower quantity and quality of crops (**Tables 6** and 7, **Figures 4–6**). The revenue reduction would be most affected by the drop in crop produce value for the three-year rotation, as it would decrease on average by 71% (V1), 69% (V2), and 61% (V3) annually (less arable fields). Subsidy payments, on the other hand, would increase by 47% (V1, V2) and 55% (V3) due to the possibility of including extensively managed meadows in the AEM scheme for organic farming (OA). The AEM payments for the preservation of butterfly grassland habitats (BH), which can be enforced under the current scheme, were not considered, as the planned management measures of alternative habitats conflict with the requirements of the AEM scheme of the Rural Development Plan.

Variable costs would be reduced by 69% (V1), 68% (V2), and 60% (V3) due to extensive land management with only one or two late grass cuts (**Tables 6** and 7). As a result, the total study area gross margin is estimated at +5% (V1), + 8% (V2), and + 15% (V3) in favor of the planned alternative habitats. By adding average AEM payments (GAL, ROT, SLP) for existing land use, as farmers applied for, the difference between gross margins are estimated at - 10% (V1), – 8% (V2), and - 1% (V3) for the planned alternative habitats in the study area. It is important to note that the gross margin per hectare of the study area would also decrease from −4 to −12%.

The planned establishment of alternative habitats will significantly impact agricultural holdings production by reducing the quantity and quality of the crop yield. Thus,

Land use	Are		Revenue (EUR)		Variable costs (EUR)		Gross margin (EUR)	
Description	**ha**	**%**	**ha**	**year**	**ha**	**year**	**ha**	**year**
Existing land use (ExU)								
Arable—conventional	92.3	64.1	1424	131,411	1040	96,029	383	35,382
Bog meadow	37.4	26.0	781	29,230	240	8987	541	20,243
Forest tree plantation	14.2	9.9	266	3785	140	1992	126	1793
Total	144	100	1142	164,425	743	107,008	399	57,418
Variant 1 (V1)—intended alternative land use								
Bog meadow	172	100	546	93,932	195	33,569	351	60,363
Total	172	100	546	93,932	195	33,569	351	60,363
Difference = V1 - ExU	+28		−596	−70,494	−548	−73,439	−48	+2945
Percentage change (%)	+19		−52	−43	−74	−69	−12	+5
Varianta 2 (V2)—intended alternative land use								
Bog meadow	172	100	559	96,204	200	34,410	359	61,794
Total	172	100	559	96,204	200	34,410	359	61,794
Difference = V2 - ExU	+28		−583	−68,222	−543	−72,598	−40	+4376
Percentage change (%)	+19		−51	−41	−73	−68	−10	+8
Varianta 3 (V3)—intended alternative land use								
Arable—organic	11	6.3	1.855	20,035	1031	11,139	824	8896
Bog meadow	161	93.8	554	89,180	198	31,884	356	57,295
Total	172	100	635	109,216	250	43,023	385	66,192
Difference = V3 - ExU	+28		−507	−55,210	−493	−63,984	−14	+8774
Percentage change (%)	+19		−44	−34	−66	−60	−4	+15

Table 6.
Calculation of the change in revenue for agriculture due to the establishment of alternative water retention habitats in the study area.

the value of produce on agricultural holdings decreases from –43 to –94% for V1, –38–94% for V2, and −21 to −94% for V3 (**Table 7**, **Figures 4–6**). Subsidy payments per individual agricultural holdings may change depending on the type of existing agricultural land management and the spatial placement of panned ecological elements of alternative habitats (marsh meadows, flood water depressions, shrub hedges, riparian vegetation on drainage ditches, and forest). Thus, all proposed variants (V1, V2, V3) range from an increase of +46% to a decrease of –59%. Revenues are also strongly negative for all agricultural holdings under all proposed variants. This trend also applies to the variable costs of all variants, as they are significantly lower than the existing ones (down to –93%). This is understandable since extensive use of marsh meadows results only in costs for grass cutting and transport. It is also interesting that the gross margins of all agricultural holdings in V1 and V2, except for one, are below negative. This means that the existing land use turns out to be more economically profitable. In V3, five farms have a gross margin higher under proposed alternative habitats. Higher gross margins resulting from the V3 envisage approximately 11 ha of

Farm ID	Alternative water retention habitat (% change from existing land use)														
	Variant 1					Variant 2					Variant 3				
	VP	SP	R	VC	GM	VP	SP	R	VC	GM	VP	SP	R	VC	GM
New agri. Land	−26	214	38	8	63	−24	210	38	8	64	−19	208	42	15	64
1	−75	−15	−61	−81	−7	−75	−15	−61	−81	−7	−42	21	−27	−50	36
2	−70	43	−37	−27	−41	−70	43	−37	−27	−41	−71	43	−37	−27	−42
3	−56	37	−29	−17	−34	−56	37	−29	−17	−34	−56	37	−29	−17	−34
4	−69	46	−35	−25	−40	−69	46	−35	−25	−40	−69	46	−35	−25	−40
5	−74	−13	−60	−80	−4	−74	−13	−60	−80	−4	−24	41	−9	−34	59
6	−83	2	−63	−81	−19	−83	2	−63	−81	−19	−83	2	−63	−81	−19
7	−54	11	−35	−24	−40	−54	11	−35	−24	−40	−54	11	−35	−24	−40
8	−85	1	−65	−83	−16	−85	1	−65	−83	−16	−85	1	−65	−83	−16
9	−94	−59	−86	−93	−65	−94	−59	−86	−93	−65	−94	−59	−86	−93	−65
10	−73	−12	−59	−80	−3	−73	−12	−59	−80	−3	−40	26	−25	−49	41
11	−43	*	46	−1	97	−38	*	58	7	114	−21	*	81	37	131
12	−71	42	−37	−28	−42	−71	42	−37	−28	−42	−71	42	−37	−27	−42
13	−76	−20	−63	−82	−12	−76	−20	−63	−82	−12	−76	−20	−63	−82	−12
14	−85	−2	−66	−83	−18	−85	−1	−65	−83	−17	−85	−1	−65	−83	−17
15	−77	−6	−60	−81	−6	−77	−6	−60	−81	−6	−63	6	−47	−68	10
16	−70	43	−37	−27	−41	−70	43	−37	−27	−41	−71	42	−37	−27	−42
17	−71	39	−39	−29	−43	−71	39	−39	−29	−43	−71	39	−39	−29	−43
18	−85	−4	−66	−83	−21	−82	−5	−64	−82	−15	−81	1	−62	−81	−11
19	−87	−26	−72	−87	−34	−83	−27	−70	−85	−28	−83	−27	−70	−85	−28
Avg.	**−71**	**47**	**−43**	**−69**	**5**	**−69**	**47**	**−41**	**−68**	**8**	**−61**	**55**	**−34**	**−60**	**15**

**no subsidy payments in existing use/increase in revenue due to subsidy payments.*
VP –Value of the Produce; SP: Subsidy Payments; R: Revenue; VC: Variable Costs; GM: Gross Margin; Avg.: average for the study area.

Table 7.
Calculation of the change in economic result (%) for agricultural holdings due to the establishment of alternative water retention habitats in the study area.

organically managed fields, with higher subsidy payments, a better selling price, and thus higher revenues.

It is important to note that 38 ha of study areas are not included in the system of subsidy payments due to their existing land use (uncultivated, overgrown). However, it represents a great potential for obtaining subsidy payments according to their existing and planned alternative use, thus having a considerable impact on the final economic calculation of the study area's gross margin (**Table 7**). Establishing alternative water retention habitats envisages 70% of wet marsh meadows, which are currently largely overgrown. In the entire study area, 49% of the land is currently in use as arable land or marsh meadow, which means that 21% of the study area needs to be cleared of overgrowth. This new agricultural land will be ready for extensive

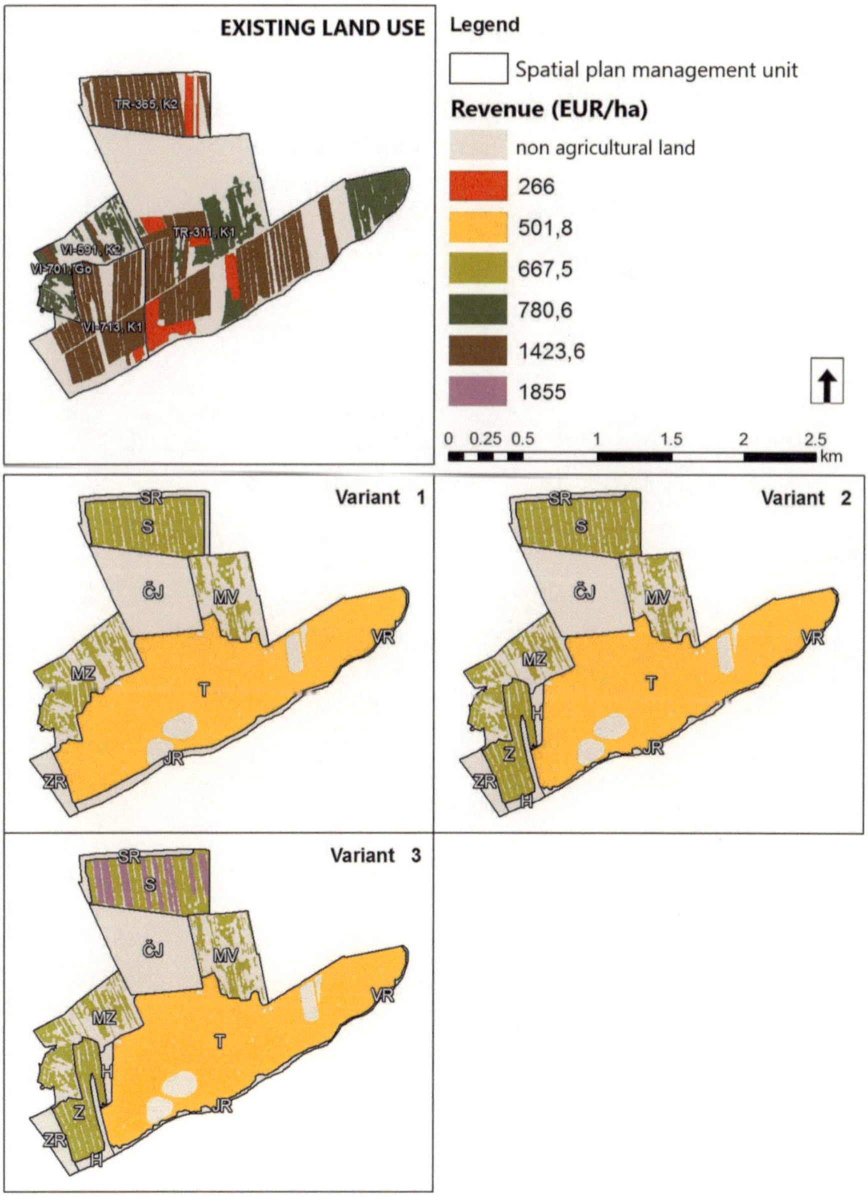

Figure 4.
Average revenue (value of the produce + subsidy payments) from agricultural land for the existing situation and variants of the alternative water retention habitats by individual spatial plan units in the study area.

management of marsh meadows after establishing alternative habitats and will thus be entitled to subsidy payments.

3.3 The impact of the alternative habitats on agriculture holdings

The planned alternative habitats will also affect individual agricultural holdings by reducing available land for production due to the establishment of permanent surface water bodies for floodwaters (depressions). In this way, some farms will be deprived

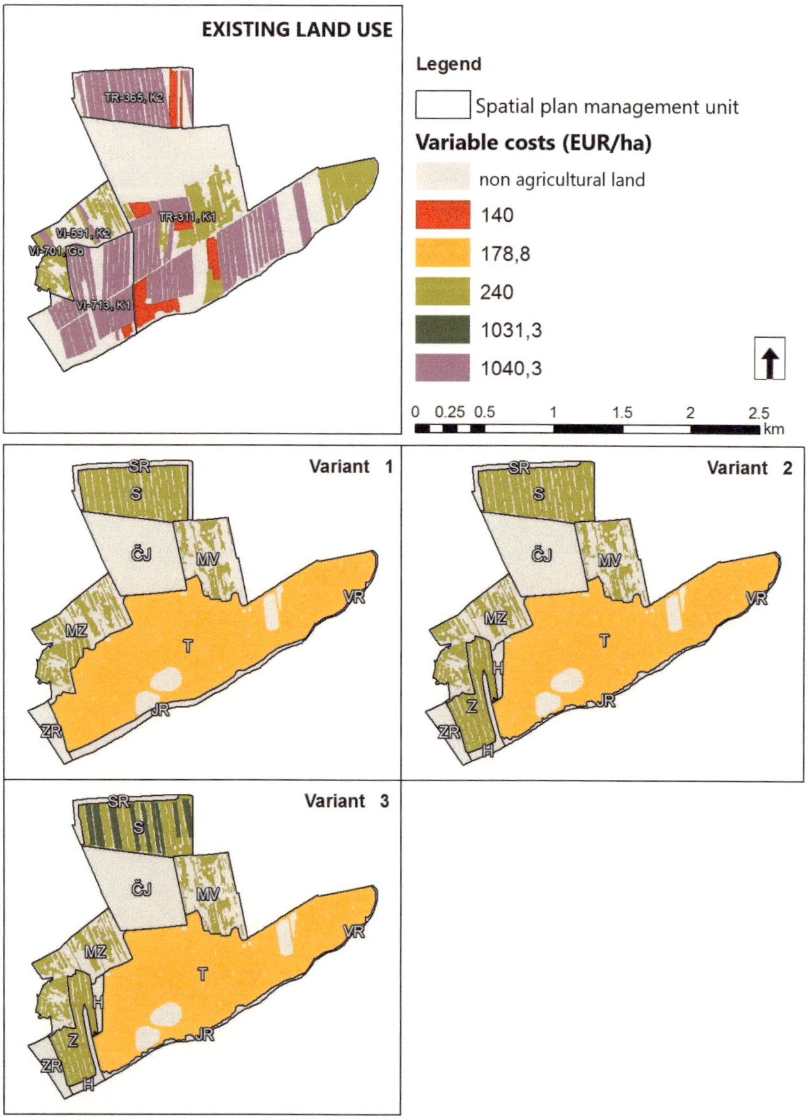

Figure 5.
Average variable costs from agricultural land for the existing situation and variants of the alternative water retention habitats by individual spatial plan units in the study area.

of practically all the agricultural land they cultivate in the area, while others will no longer be economically justifiable to cultivate the land. Such sharp interventions in the production scale can significantly impact the individual agricultural holding socioeconomic situation. Furthermore, less fodder production also leads to a reduction in livestock production.

Variants V1 and V2 with floodwater depressions are the same in terms of their impact on agricultural holdings, as they envisage the permanent occupation of 8.9 ha of agricultural land of which 5.6 ha are arable land, 2.3 ha are plantations of forest trees, and 1 ha of overgrown land. Thus, the total decrease of gross margin for V1 and V2 due to floodwater depressions is 2452 EUR annually without considering any of the

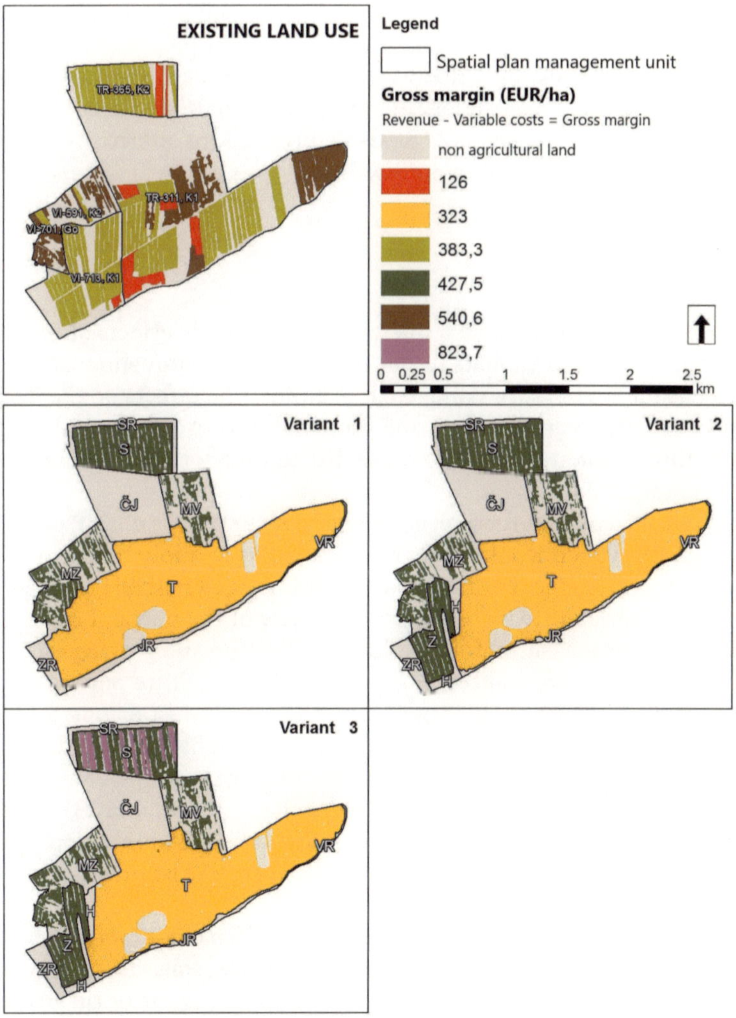

Figure 6.
Average gross margin (revenue—Variable costs) from agricultural land for the existing situation and variants of the alternative water retention habitats by individual spatial plan units in the study area.

AEM payments for arable lands. On the other hand, the V3 also envisages the permanent occupation of 8.9 ha of agricultural land, of which 3.4 ha is arable land, 2.3 ha is plantations of forest trees, and 2.7 ha is overgrown land. Thus, the decrease of gross margin for V3 would be lower - 1619 EUR annually without considering possible AEM payments for arable lands.

Seven farms will be affected to a certain degree by permanent land loss due to floodwater depressions. According to variants V1 and V2, the most affected agricultural holding No. 18 would permanently lose 2.24 ha of arable land, while according to variant V3, the loss would be minimal with only 0,09 ha of arable land. Agricultural holdings No. 9 and 19 that cultivate arable land would permanently lose 1.56 and 1.36 ha of fields, respectively, according to all three variants of the alternative habitats. The agricultural holding No. 11, which is engaged in producing wood on plantations of forest trees, would lose 2.24 ha of land according to all three variants.

In order to avoid the permanent destruction of the soil profile and potential for agricultural production, we suggest that the placement of permanent measures destroying agricultural soils (e.g., floodwater depressions, sedimentation basin) should avoid agricultural areas under cultivation with high production potential. Furthermore, given the extensiveness of areas intended for alternative habitats, they should be constructed in areas of overgrowth or forests.

4. Conclusions

This research is unique because it analyses the economic effects of establishing alternative water retention habitats in the area of existing conventional crop production on agricultural holdings. In doing so, it examines the effects on the value of the crop produce, revenue, variable costs and gross margins of agricultural holdings, and the effects of three variants of new land use distribution within the framework of alternative habitat establishment.

The economic calculation for alternative habitats includes subsidy payments for organic agriculture, which would make the most sense to apply for in an area with extensive use. However, the question is whether it is even possible to integrate these areas into organic agriculture from the point of view of the nutrient cycle since the planned management measures do not allow fertilization and grazing only on a small scale. Therefore, if we wanted to meet organic agriculture requirements, we would need 0.2 livestock units (LU) per hectare for the lowest payment for organic grasslands and 0.5 LU per ha for a higher payment on arable land. In the study area, where 172 ha of extensive wetland meadows are planned, this means either 35 or 86 LU, equal to the same number of cows (older than 2 years) and 233 or 573 sheep (older than 1 year), respectively. In case it would not be possible to apply for organic farming payments farming in the study area would be practically unprofitable.

The classification species Eurasian woodcock (*Scolopax rusticola*) in the study area requires special living conditions (grassland with soft wet soils and riparian vegetation on the edge of meadows). Due to that, the management regime of the planned alternative habitats (grass cutting at different dates) prevents the enforcement of AEM scheme subsidy payments from the Rural Development Programme for measures BH (butterfly grassland habitats) and STE (grassland cut in late summer), which further reduces the gross margin of alternative habitats variants. Furthermore, according to the spatial databases, the study area is preferentially protected for butterfly grassland habitats (BH) and grassland cut in late summer (STE) and not for the conservation of bird habitats of extensive wet grasslands (VTR) in Natura 2000 areas. In this case, the management defined in the Municipal Spatial Plan for the alternative habitats in the study area and the official State databases for protecting species do not match. Therefore, there are two options to preserve agriculture in this area: (i) a change in management measures in the Spatial Plan or (ii) designation of this area as a habitat for birds of extensive wet grasslands under Natura 2000.

It is expected that after establishing alternative habitats, most agricultural holdings renting or leasing agricultural land will cancel contracts with land owners. At the same time, we would like to point out that agricultural holdings farming agricultural land leased by the State Agricultural Land and Forest Fund (Fund) have the right to farm till the end of the lease contract, usually 10 years. However, Fund has the right to change the contract at any time.

When establishing alternative habitats, care must be taken that land use or management conditions do not affect the agriculture holdings contract with the state for the implementation of the AEM (organic farming, butterfly grassland habitats grassland cut in late summer, bird habitats of extensive wet grassland, and greening), which usually expires after 5 years. If the land use or land management changes or the lease contract is to be terminated earlier, it is necessary to provide the farmer with replacement land, where it is possible to implement the AEM for which the contract was signed. Otherwise, the reduction of the land area included in the AEM scheme could constitute a breach of the contract, and the agricultural holding would be sanctioned with a reduction of subsidy payments by a certain percentage.

Considering the planned management measures of alternative habitats envisaged in the study area, it can be argued that there are no development opportunities for conventional, integrated, or organic farming. The existence of any agricultural activity after the establishment of habitats in the area will depend entirely on subsidy payments. Due to the management requirements, subsidy payments would be significantly reduced by excluding AEM payment due to the different goals pursued in the study area by the Ministries responsible for agriculture and environment (butterfly habitats, meadow habitats) and Municipal Spatial Plan of the Municipality of Ljubljana (habitats for birds of wet meadows).

The only future development opportunity for agricultural holdings in the study area is providing ecosystem services for grass-cutting marsh meadows. Depending on the size of the study area and the number of cuts, it would be sufficient if one or two farms would provide their services. However, with such a management regime, it will be necessary to find additional funds to pay for the service.

Compared to the existing use, the arrangement of alternative water retention habitats with wet marsh meadows pursues an entirely different goal: the establishment of a habitat for classification species, especially the Eurasian woodcock (*S. rusticola*), which needs specific conditions for its survival. Furthermore, a nature protection goal differs from an agricultural one, which follows the provision of agricultural land for the economically justifiable performance of the agricultural activity and the provision of food and jobs. Thus, most of the development possibilities are in the local, green, organic, hiking, cycling, and photo tourism, which is not the primary domain of agriculture but offers new development opportunities in the diversification of activities on the agricultural holdings.

Acknowledgements

This chapter was prepared based on the Expert Report on the Analysis of the Impacts of the Establishment of Alternative Habitats on Agriculture for the OPPN 398 Study Area (430-1161/2012 ZIN-31/12), supported by the City Municipality of Ljubljana (Mestna občina Ljubljana).

Author details

Matjaž Glavan
Biotechnical Faculty, University of Ljubljana, Ljubljana, Slovenia

*Address all correspondence to: matjaz.glavan@bf.uni-lj.si

References

[1] Lankoski J. Alternative Payment Approaches for Biodiversity Conservation in Agriculture. Paris: OECD Food, Agriculture and Fisheries Papers, No. 93, OECD Publishing; 2016. DOI: 10.1787/5jm22p4ptg33-en

[2] Klemenčič T, Kink B. Nadomestni habitat - omilitveni ali izravnalni ukrep? = replacement habitat - mitigation or compensatory measure? Varstvo narave. 2015;**28**:27-40

[3] Jančar T, Govedič M. Variantni Predlogi Ureditve Nadomestnih Habitatov Za širitev Ljubljanskega smetišča = Variant Proposals for Arranging Alternative Habitats Due to the Expansion of the Ljubljana Landfill. Ljubljana: Društvo za opazovanje in preučevanje ptic Slovenije – BirdLife Slovenia; 2013. p. 2013

[4] Fazarinc R. Hidravlična študija širitve Deponije Barje in Vzpostavitve Nadomestnih Habitatov = Hydraulic Study of the Expansion of the Barje Landfill and the Establishment of Alternative Habitats. Ljubljana: IZVO-R; 2013

[5] Fridl J, Perko D, Orožen Adamič M, Belec B, Brancelj A, Gabrovec A, et al. Slovenija – pokrajine in ljudje = Slovenia - landscapes and people. Ljubljana: Mladinska knjiga; 1990. p. 735

[6] Jernej S. Analiza Mesta Ljubljane z Dodatkom Aplikacij Za načrtovanje Klimatskih izhodišč Pri načrtovanju Rabe Prostora = Analysis of the Climate of the City of Ljubljana with the Addition of Applications for Planning Climatic Starting Points When Planning the Use of Space. Gradec: Geography Institute Karl-Franzens University; 2000. p. 252

[7] Urbančič M, Simončič P, Prus T, Kutnar L. Atlas Gozdnih Tal Slovenije = Atlas of Forest Soils of Slovenia. Ljubljana: Zveza gozdarskih društev Slovenije, Gozdarski vestnik, Gozdarski inštitut Slovenije; 2005. p. 100

[8] Jerič D, Caf A, Demšar-Benedečič A, Leskovar S, Oblak O, Soršak A, et al. Katalog kalkulacij za načrtovanje gospodarjenja na kmetijah v Sloveniji = Catalog of calculations for farm management planning in Slovenia. Ljubljana: Kmetijsko gozdarska zbornica Slovenije – Chamber of Agriculture and Forestry of Slovenija; 2011 266 p

Chapter 5

Managing Prior Converted Hydric Soils to Support Agriculture Production and Maintain Ecosystem Services: A Dedicated Outreach to the Agriculture Community

Michael Aide, Samantha Siemers Indi Braden, Sven Svenson, Shakirah Nakasagga, Kevin Sargent, Miriam Snider and Marissa Wilson

Abstract

Hydric soils and prior converted soils are frequently used for agricultural production. Agriculture production and their associated agribusinesses are the chief economic sector; thus, agriculture is critical for rural prosperity. However, the continuous production of grain crops increases the risk of disease and insect outbreaks, which may lead to soil nutrient exhaustion or substantial usage of annual fertilizer amendments, loss of soil carbon, and soil structure degradation attributed primarily to tillage, decrease in biodiversity, and increased soil compaction. At the David M. Barton Agriculture Research Center at Southeast Missouri State University, our focus has been to support profitable agriculture production and environmental stewardship. We have developed a decade-long research program specializing in subsurface controlled irrigation with the gradual development of edge-of-field technologies. We further developed a constructed wetland to address nutrient pollution concerns with confined feeding operations. Pastures associated with the confined feed facility and the constructed wetland have initiated a soil health program. Our evolution has now permitted the David M. Barton Agriculture Research Center to become a regional center to showcase the relationships that support both profitable agriculture and environmental stewardship.

Keywords: prior converted wetlands, subsurface drainage, denitrification bioreactors, constructed wetlands, soil health

1. Introduction

Knowledge of water and nutrient flux in wetlands is integral to land management across southeastern Missouri. The region has the largest completed land drainage project in the USA [1]. The Little River Drainage Project converted 1.6 million ha (4 million acres) of marshlands into productive croplands. The economic development of the region is primarily vested in agriculture; however, the realization that the restoration of ecosystem services is important for water quality, soil health, nutrient management, habitat preservation, and advancing biological diversity is emerging. This vast region currently produces corn (*Zea mays*), soybeans (*Glycine max*), wheat (*Triticum aestivum*), rice (*Oryza sativa*), cotton (*Gossypium hirsutum*), and specialty crops. Livestock includes beef (*Bos taurus*), swine (*Sus domesticus*), sheep (*Ovis aries*), and chicken (*Gallus gallus domesticus*).

Nitrogen migration from croplands supports eutrophication of freshwater resources and results in hypoxia across the Louisiana and Texas continental shelf [2, 3]. Additionally, the United States Environmental Protection Agency established 10 mg NO_3-N L^{-1} as the nitrate drinking water standard; however, 1.5 mg NO_3-N L^{-1} may support eutrophication [2]. A significant portion of the Mississippi River nitrate discharge into the Gulf of Mexico is derived from 15 million ha of artificial drainage within the Mississippi River watershed [1, 2]. Aide et al. [3] demonstrated that the nitrate concentrations from tile drainage effluents were a function of rainfall after nitrogen fertilization involving corn. Soil analysis demonstrated that nitrate was effectively leached to the tile-drainage technology.

The objective of this article is to demonstrate how to develop and install infrastructure that supports both production agriculture and environmental stewardship.

2. Research to limit Nitrogen transport from tile-drained agricultural lands

Tile drainage is common across the US corn belt, providing removal of excess water. Much of the drainage is uncontrolled, implying that the producer may not have the capacity to limit the tile drainage. Advantages of tile drainage include: (i) creating soil aeration permitting optimal root and seed respiration; (ii) promoting soil warming, especially in the spring; (iii) timely field operations; and (iv) minimizing nitrogen loss because of denitrification. A key disadvantage of tile drainage is the leaching losses of nitrate and sulfate, which require additional fertilization and threaten water quality [4–7]. Faust et al. [7] evaluated management practices used in drainage ditches to reduce (i) total suspended solids and (ii) nitrogen and phosphorus concentrations, especially for moderate rainfall intensities.

Agronomic approaches to limiting nitrogen losses from tile-drainage fields include: (i) appropriate the timing and rates of nitrogen fertilizers, (ii) anticipate the nitrogen supply arising from mineralization, (iii) establish appropriate yield goals, (iv) utilize urease and nitrification inhibitors, (v) monitor crop nutrient status, (vi) employ diverse crop rotations and implement cover crops, (vii) manage plant residues, (viii) utilize precision fertilization practices, and (ix) install riparian buffers and other edge-of-field technologies [8].

2.1 Edge-of-field technologies to limit Nitrate degradation of water resources

Aide et al. [2, 3, 8] discussed the installation and evaluation of edge-of-field technologies primarily engineered to eliminate nutrient transport from croplands. Aide et al. [3] demonstrated that a denitrification bioreactor effectively reduced nitrate-N concentrations from 69 mg NO_3-N L^{-1} to 21 mg NO_3-N L^{-1} from May through June (2015). For the 2018 corn harvest, Aide et al. [4] reported that the mean tile-drainage nitrate concentration ranged from 1.5 to 109 mg NO_3-N L^{-1}. The influent drainage into the denitrification bioreactor ranged from 0.4 to 103 mg NO_3-N L^{-1}, whereas the outlet drainage from the denitrification bioreactor ranged from 0.3 to 5.2 NO_3-N L^{-1}. The smaller tile-drainage nitrate concentrations in 2019 were approximately 1.6 to 4.5 mg NO_3-N L^{-1} because of soybean cultivation and the lack of nitrogen fertilization. Data for subsequent years corroborates the presented findings.

2.2 Constructed wetlands to capture nutrient-laden overland flow

Constructed wetlands are engineered soil infrastructures designed to capture overland flow and subsequently facilitate soil-vegetation pathways to convert water-bearing nutrients into plant materials. Constructed wetlands enhance ecosystems by enhancing hydrological, biological, geochemical, and pedogenic processes that improve water quality and other ecosystem services. Perceived advantages of constructed wetlands include: (i) on-site nitrogen and phosphorus conversions into plant materials, (ii) reduced biological and chemical oxygen demands, (iii) odor reduction, (iv) wildlife habitat, (v) esthetics, and (vi) potential economic benefits [9–16].

2.3 Cover crops

Cover crops are used primarily to (i) constrain wind and water erosion, (ii) enhance available water capacity, (iii) suppress weeds and reduce herbicide usage, (iv) become compatible with an integrative pest management system to limit the incidence of specific insect and pathogens, (v) augment soil porosity and maintain appropriate soil bulk densities, (vi) convert soil nitrate and phosphate to plant-based organic nitrogen and phosphate to reduce off-site nutrient migration, and (vii) increase soil organic matter contents. The choice of plant speciation of the cover crop annually is governed by crop rotation, soil nutrient concentrations, and economics concerning seed purchase. Wheat (*T. aestivum*) and rye (*Secale cereale*) are popular cover crop choices, frequently interseeded with forage legumes.

2.4 Soil health and pasture management

Proper rotational grazing is integral to maintaining a vibrant forage program. However, for most producers, forage production detractions occur because of weather, forage species competitiveness, weed and disease management, soil fertility programs, the intensity and oversight of the rotational grazing program, and other factors. The United Sates Department of Agriculture—Natural Resources and Conservation Service defines soil health as follows: “Soil health is defined as the continued capacity of soil to function as a vital living ecosystem that sustains plants, animals, and humans” [17]. Soil health provides five key services: (i) regulating water, (ii) sustaining plant and animal life, (iii) filtering and buffering potential pollutants, (iv) cycling nutrients, and (v) providing physical stability and support. Landowner management may support soil health by (i) maximizing the presence of living roots, (ii) minimizing the disturbance because of tillage and animal traffic, (iii) maximizing soil cover with living plant material, and (iv) maximizing biodiversity [17].

Soil quality is assessed individually for each soil and is documented and measured using indicators [18–26]. The relevant indicators in pastures that we employ to document soil health improvements include: (i) physical attributes (rooting depth, bulk density, and infiltrate capacity), (ii) chemical attributes (total organic carbon, total organic nitrogen, labile (active) carbon, and pH), and (iii) biological attributes (microbial carbon biomass, microbial nitrogen biomass, potential N mineralization, phospholipid fatty acids, and soil respiration).

3. Existing infrastructure at the David M. Barton agriculture research center to support profitable production agriculture and environmental sustainability

Southeast Missouri State University is a regional comprehensive public university that provides student-centered education and experiential learning experiences across the curriculum. The David M. Barton Agriculture Research Center, located at Cape Girardeau County (Missouri, USA), is an experiential learning facility for the Department of Agriculture at Southeast Missouri State University. **Figure 1** illustrates the spatial distribution of the environmental technologies and the material transport pathways.

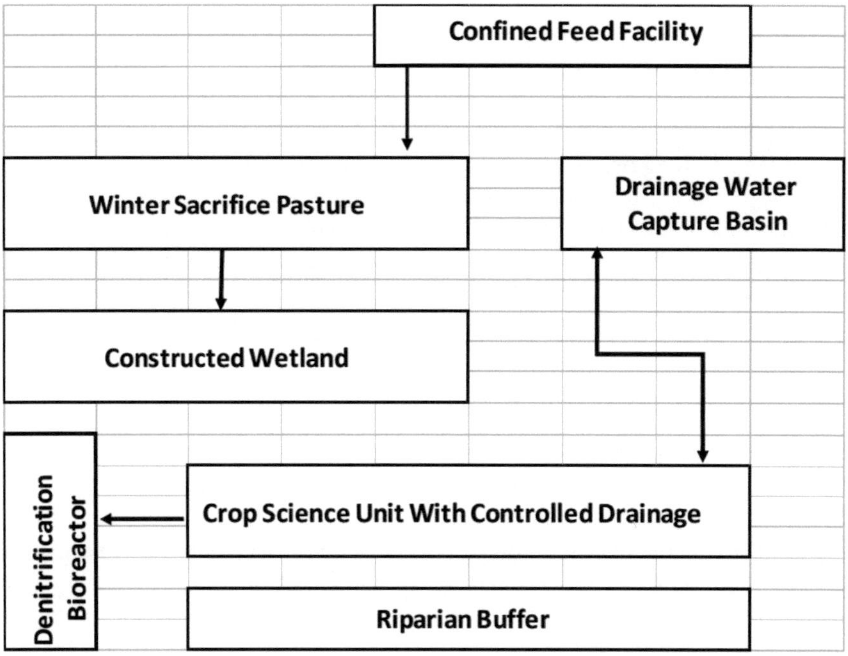

Figure 1.
Map of the infrastructure layout.

3.1 Study area climate

The mean annual temperature is approximately 13°C (56°F), and the mean annual precipitation is approximately 1.12 mm (44 inches) [27]. The mean monthly temperature for January is 3°C, and the mean monthly temperature for July is 25°C. Peak temperatures typically occur in July, with some days having a maximum near 40°C (104°F). Rainfall is typically greater from March to May; however, Gulf of Mexico weather events may provide heavy rain events from June to October. The mean October rainfall is 7 cm, whereas the mean May rainfall is 13 cm. The growing season is approximately 210 days [27].

3.2 The soil resource

The Wilbur series (coarse silty, mixed, superactive, mesic Fluvaquentic Eutrudepts) is the dominant soil series in the Crop Science Unit (Bottomlands). The pedons are very deep, moderately well-drained, permeable soils formed in silt loam alluvium that display an Ap–Bw–Cg horizon sequence. Saturated hydraulic conductivity is 4.2 to 14.1 micrometer sec^{-1}, and the permeability is moderate. The soil pH ranges from slightly acidic to neutral in the ochric epipedon and strongly acidic (pH 5.1 to 5.5) and very strongly acidic (pH 4.5 to 5.0) in the cambic and deeper soil horizons, respectively.

Upland landscapes contain soils formed in thick loess and exhibit 2 to 6 percent slopes. The Menfro series (fine silty, mixed, superactive, mesic Typic Hapludalfs) consists of very deep, well-drained, moderately permeable soils exhibiting A–E–BE–Bt horizon sequences. The Winfield series (fine silty, mixed, superactive, mesic Oxyaquic Hapludalfs) consists of very deep, moderately well-drained soils exhibiting A–E–BE–Bt–Btg horizon sequences. Both soil series have argillic horizons, exhibiting moderately acidic to strongly acidic pH levels.

3.3 Crop science infrastructure overview

The David M. Barton Agriculture Research Center has a 40 ha (100 acre) crop science unit featuring a controlled subsurface drainage and irrigation technology. The subsurface controlled drainage system design involves parallel tiles having 10-meter spacing. Irrigation and drainage are monitored and regulated by using stop-log boxes fitted with adjustable baffles to permit irrigation/drainage water to be added/removed by gravity flow. Submersible pumps support the irrigation.

A 12×10^3 $meter^3$ (3.3×10^6 gallon) tile-drainage water capture basin was constructed to store excess tile-drainage water collected during the off-season to be reapplied as subsurface irrigation water during the cropping season, thus reapplying nitrogen to support plant growth and development.

A denitrification bioreactor is connected to the controlled-subsurface irrigation and drainage technology to receive drainage effluent. The denitrification bioreactor was designed and installed to transform nitrate to inert nitrogen gas (N_2), nitric oxide (NO), or nitrous oxide (N_2O). The relative speciation of nitrate-N into the three nitrogen gaseous species is pH dependent. Notably, in spring and summer rainfall events, the denitrification bioreactor consistently receives tile-drainage influents having nitrate-N concentrations between 20 and 40 mg NO_3-N L^{-1} and having effluent discharges from 3 to 10 mg NO_3-N L^{-1} [2, 3].

A riparian buffer is an edge-of-field technology designed to limit nutrient-laden runoff from entering freshwater resources. The riparian buffer is designed as 22.9 meters (75 ft) of trees and understory, with 7.6 meters (25 ft) of warm-season grasses. The riparian buffer is along an order III stream, and all trees, shrubs, and grasses/forbs are native. Collectively, the riparian buffer and the denitrification bioreactors are designed to limit nutrient migration from the crop production area to freshwater resources.

3.4 Animal science infrastructure overview

The animal science unit primarily focuses on cow-calf production with dedicated infrastructures including: (i) a pavilion for animal care and breeding, (ii) a semi-confined feed facility, and (iii) a confined feed facility. A grazing paddock system consists of 56 ha (140 acres) primarily having cool-season hay/pastures (tall fescue or *Schedonorus arundinaceus*) and warm-season grass pastures (bermudagrass or *Cynodon* spp). Water is provided through underground conduit that is fitted with freeze-preventive hydrants.

4. Research involving agriculture production and environmental Stewardship

4.1 Crop production

The Crop Science Unit maintains a corn (*Z. mays*) and soybean (*G. max*) rotation. Research involving the corn–soybean rotation is conducted annually to better estimate the influence of agronomic practices on the concentrations of tile drainage nitrate. Research involving nitrate tile drain concentration variations were attributed to: (i) nitrogen fertilization timing and rates; (ii) nutrient uptake patterns over crop growth stages, harvest removal, and residue return; and (iii) crop yields and their contribution to farm profitability.

For the 2022 harvest season, soybean yields were spatially variable but averaged from 4036 kg ha^{-1} (60 bushels $acre^{-1}$) when planted after wheat and 4372 kg ha^{-1} (65 bushels $acre^{-1}$) for full season (planted after cover crop). For the 2021 and 2022 growing seasons, we estimated harvest loss and residue return for nitrogen, phosphorus, potassium, sulfur, magnesium, and calcium (**Table 1**).

The data simply illustrates quantitative assessment of nutrient cycle components that are integral to assessing land management influences. Note that harvest removal and residue return concentrations influence soil fertility, the potential for nutrient leaching and water quality, soil microbial activity, and wildlife habitat.

	Nitrogen	Phosphorus	Potassium	Sulfur
Harvest removal	225	25	73	14
Residue return	40	7	17	5

Table 1.
Harvest removal and residue return (kg ha^{-1}) for key nutrients for 2021 soybean.

4.2 Manure nutrient capture zones and a constructed wetland to inhibit Nitrogen and Phosphorus flux

In 2022, we installed a land-graded constructed wetland to provide discrete zones having different water saturation intensities and durations. Nutrient bearing inflow into the constructed wetland occurs from the winter sacrifice pasture. Water overland flow is channeled by a terrace system. In spring 2023 we will seed native aquatic plants to document which plant species are most suited to the constructed wetland and its difference water saturation zones.

The research objectives for the constructed wetland must be visioned with the manure-laden winter sacrifice pasture and the associated confined feed facility. Our objectives are: (i) to evaluate a constructed wetland to reduce nitrogen and phosphorus transport and impact to freshwater resources, (ii) to assess the aquatic plant composition for augmenting ecosystem services and compatibility across different water saturation regimes, and (iii) to determine if selected aquatic plants may be harvested for resale. Associated with the constructed wetland is a series of grazing pastures. Our soil health program is designed to merge the benefits of soil health with advanced grazing practices [28].

4.3 Connectivity of environmental Stewardship and farm profitability to support producer acceptance

Wetlands provide benefits, including: (i) critical habitat and breeding grounds, (ii) feeding and resting grounds for migratory birds and habitat corridors, (iii) recreational and esthetic benefits, (iv) reduction of erosion and flooding, (v) moderation of groundwater levels and base flow, (vi) assimilation of nutrients, and (vii) protection of drinking water sources [29]. Expertly managed upland pastures also provide benefits, including: (i) forage for livestock, (ii) supporting rainfall infiltration and

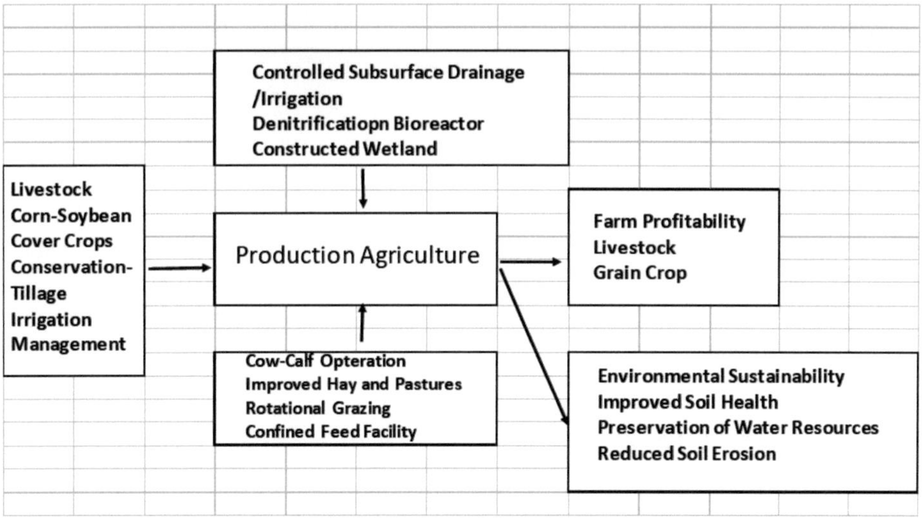

Figure 2.
Illustration for modeling information flow.

reducing overland flow to nearby streams, (iii) with vigorous vegetation growth encouraging nutrient cycling, (iv) reducing the quantity of fertilizer amendments, (v) distributing manure across a greater area, (vi) increasing carbon sequestration levels, and (vii) augmenting farm profitability.

Our outreach goal is to provide meaningful and informative learning activities to a diverse audience, wherein we concentrate on farm profitability and environmental sustainability. The outreach programing focuses on aligning agricultural production with viable and environmental-based cultural practices and incorporating applicable soil engineering structures (**Figure 2**). The topics that the faculty address to the agricultural community include: (i) controlled subsurface drainage/irrigation, (ii) edge-of-field technologies, (iii) modern pasture management, (iv) soil health, and (v) agronomic practices to augment economic and sustainable crop yields. Audiences include a single producer to producer workshops, agriculture educators and their students, and state and federal personnel. Social media is being developed for more distant interested individuals.

5. Conclusion

The purpose of this article is to demonstrate how to develop and install infrastructure that supports both production agriculture and environmental stewardship. At the David M. Barton Agriculture Research Center, the infrastructure development and installation include: (i) a controlled subsurface drainage and irrigation technology, (ii) a denitrification bioreactor to limit tile-drainage nitrate concentrations, (iii) riparian corridors, (iv) a drainage water capture basin to reuse drainage water for irrigation, and (v) a constructed wetland and a confined beef feeding facility. Collectively, these infrastructures permit the teaching and outreach capabilities to link production agriculture and environmental stewardship.

Author details

Michael Aide*, Samantha Siemers Indi Braden, Sven Svenson, Shakirah Nakasagga, Kevin Sargent, Miriam Snider and Marissa Wilson
Department of Agriculture, Southeast Missouri State University, Cape Girardeau, USA

*Address all correspondence to: mtaide@semo.edu

References

[1] Aide MT, Aide C, Braden IS, McVay B. A large-scale wetland conversion project in southeastern Missouri: Sustainability of water and soil. In: Gokce D, editor. Wetlands Management: Assessing Risks and Sustainable Solutions. Rijeka, Croatia: InTech; 2018. DOI: 10.5772/Interchopen.81254

[2] Aide MT, Braden IS, Siemers S, Svenson S. Soil water infrastructure to eliminate off-site nutrient migration and support farm profitability. Agricultural Sciences. 2022;**13**:776-789. DOI: 10.4236/as.2022.136050

[3] Aide MT, Braden IS, Svenson S. Edge of field technology to eliminate nutrient transport from croplands: Specific focus on denitrification bioreactors. In: Larramendy ML, Soloneski S, editor. Soil Contamination. Rijeka, Croatia: InTech; 2016. pp. 3-21. ISBN 978-953-51-2816-8. DOI: 10.5772/64602

[4] Dinnes DL, Karlen DL, Jaynes DB, Kaspar TC, Hatfield JL, Colvin TS, et al. Review and interpretation: Nitrogen management strategies to reduce nitrate leaching in tile-drained Midwestern soils. Publications from USDA-ARS/UNL Faculty.263. 2002. Available from: https://digitalcommons.unl.edu/usdaarsfacpub/263

[5] Arenas Amado A, Schilling K, Jones C, Thomas N, Weber L. Estimation of tile drainage contribution to streamflow and nutrient loads at the watershed scale based on continuously monitored data. Environmental Monitoring and Assessment. 2017;**189**(9):486. DOI: 10.1007/s10661-017-6139-4

[6] Lenhart C, Gordon B, Gamble J, Current D, Ross N, Herring L, et al. Design and hydraulic performance of a tile drainage treatment wetland in Minnesota, USA. Water. 2016;**8**(12):549-568. DOI: 10.3390/w8120549

[7] Faust DR, Kröger R, Moore MT, Rush SA. Management practices used in agricultural drainage ditches to reduce Gulf of Mexico Hypoxia. Bulletin of Environmental Contamination and Toxicology. 2018;**100**(1):32-40. DOI: 10.1007/s00128-017-2231-2

[8] Aide MT, Braden IS, Mauk D, McAlister D, McVay B, Murray S, et al. Tile-drain and denitrification bioreactor water chemistry for a soybean (*Glycine max* (L.) Merr.)-Corn (*Zea mays* L.) rotation in East-Central Missouri (USA). Journal of Geoscience and Environment Protection. 2020, 2020;**8**:143-154. DOI: 10.4236/gep.2020.84010

[9] Aide MT, Braden I, Svenson S, Siemers S, Murray S. Employing constructed wetlands to sustainably manage nutrient-bearing water: A review with an emphasis on soil behavior and effluent nutrient reduction. Journal of Geoscience and Environment Protection. 2020;**8**:94-106. Available from: https://www.scirp.org/journal/gep ISSN Online: 2327-4344

[10] Carvalho PN, Arias CA, Brix H. Constructed wetlands for water treatment: New developments. Water. 2017;**9**:397. DOI: 10.3390/w9060397

[11] Scholz M, Hedmark A. Constructed wetlands treating runoff contaminated with nutrients. Water Air Soil Pollution. 2010;**205**:323-332. DOI: 10.1007/s11270-009-0076-y

[12] Thalla AK, Devatha CP, Anagh K, Sony E. Performance evaluation of horizontal and vertical flow constructed

wetlands as tertiary treatment option for secondary effluents. Applied Water Science. 2019;**9**:147. DOI: 10.1007/s13201-019-1014-9

[13] Risko A, Ramani B, Seyoum L, Helmut K. Nitrogen removal in integrated anaerobic–aerobic sequencing batch reactors and constructed wetland system: A field experimental study. Applied Water Science. 2019;**9**:136. DOI: 10.1007/s13201-019-1015-8

[14] Ali Z, Mohammad A, Riazb Y, Quraishia UM, Malik RN. Treatment efficiency of a hybrid constructed wetland system for municipal wastewater and its suitability for crop irrigation. International Journal of Phytoremediation. 2018;**20**:1152-1161. DOI: 10.1080/15226514.2018.1460311

[15] Lu S, Zhang P, Jin X, Xiang C, Gui M, Zhang J, et al. Nitrogen removal from agricultural runoff by full-scale constructed wetland in China. Hydrobiologia. 2009;**621**:115-126. DOI: 10.1007/ s10750-008-9636-1

[16] Jesus JM, Danko AS, Fiuzal A, Borges M-T. Effect of plants in constructed wetlands for organic carbon and nutrient removal: A review of experimental factors contributing to higher impact and suggestions for future guidelines. Environmental Science and Pollution Research. 2018;**25**:4149-4164. DOI: 10.1007/s11356-017-0982-2

[17] Available from: https://www.nrcs.usda.gov/conservation-basics/natural-resource-concerns/soils/soil-health (verified January 16, 2023)

[18] Aide M, Braden IS, Murray S, Schabbing C, Scott S, Siemers S, et al. Optimizing beef cow-calf grazing across Missouri with an emphasis on protecting ecosystem services. Land. 2021;**10**:1076. DOI: 10.3390/land10101076

[19] Karlen DL, Andrews SS, Wienhold BJ, Zobeck TM. Soil quality assessment: Past and future. Journal of Integrative Biosciences. 2008;**6**:3-14 Available from: https://digitalcommons.unl.edu/usdaarsfacpub/1203/

[20] Teague W. Forages and pastures symposium: Cover crops in livestock production: Whole system approach: Managing grazing to restore soil health and farm livelihoods. Journal of Animal Science. 2008;**96**(4):1519-1530

[21] Teague R, Kreuter U. Managing grazing to restore soil health, ecosystem function, and ecosystem services. Frontiers in Sustainable Food Systems. 2020;**4**:1-13. DOI: 10.3389/fsufs.2020.534187

[22] Kruse JS. Framework for sustainable soil management literature review and synthesis. Soil and Water Conservation Society. SWCS Special Publication 2007.001. 2007. Available from: https://www.joinforwater.ngo/sites/default/files/library_assets/LAN_E6_framework_sustainable.pdf

[23] Paudel BR, Udawatta RP, Kremer RJ, Anderson SH. Soil quality indicator responses to row crop, grazed pasture and agroforestry buffer management. Agroforestry Systems. 2012;**84**:311-323. DOI: 10.1007/s10457-011-9454-8

[24] Karlen DL, Obrycki JF. Measuring rotation and manure effects in an Iowa farm soil health assessment. Agronomy Journal. 2018;**110**:63-73. DOI: 10.2134/agronj2018.02.0113

[25] Dahal S, Franklin D, Subedi A, Cabrera M, Hancock D, Mahmud K, et al. Strategic grazing in beef-pastures for improved soil health and reduced runoff-nitrate. A step towards sustainability. Sustainability. 2020;**12**:558. DOI: 10.3390/su12020558

[26] Derner JD, Smart AJ, Toombs TP, Larsen D, McCulley RL, Goodwin J, et al. Soil health as a transformational change agent for US grazing lands management. Rangeland Ecology & Management. 2018;**71**:403-408. DOI: 10.1016/j.rama.2018.03.007

[27] Festervand DF. Soil survey of Cape Girardeau, Scott and Mississippi Counties, Missouri. Produced in cooperation with the United States Department of Agriculture, United States Forest Service, and the University Missouri-Columbia. Printed Washington DC. 1981

[28] Bilotta GS, Brazier RE, Haygarth PM. The impacts of grazing animals on the quality of soils, vegetation, and surface waters in intensively managed grasslands. Advances in Agronomy. 2007;**94**:237-280. DOI: 10.1016/S0065-2113(06)94006-1

[29] Daily GC, Matson PA, Vitousek PM. Ecosystem services supplied by soil. In: Nature's Services: Societal Dependence on Natural Ecosystems. Washington, DC: Daily, G.C.; Island Press; 1997. pp. 113-132

Chapter 6

Monitoring the Properties of an Abandoned Depleted Peat Bog to Determine the Prospects for Use

Anisimova Tatiana Yuryevna

Abstract

Peatlands after drainage can be effectively used as highly productive agricultural grasslands. The preservation of the fertility of peat soils depends on the nature of their use in agricultural production. Irrational and illogical use of peat bogs leads to loss of organic matter and nitrogen and reduction of their reserves. Currently, these deposits are often in the form of abandoned and overgrown forests. The appearance of disturbed landscapes leads to negative changes in vegetation and soil cover, water and temperature balance of the area, composition of soil, waste water and development of water, and wind erosion. The results of monitoring changes in some soil properties of a peat bog over a 20-year period are presented. The results of the geobotanical survey of the peat massif, which was conducted for the first time, are presented. The influence of the action of biotic and abiotic factors on the change of agrochemical characteristics of anthropogenic-transformed peat soil is determined. Depending on the degree of development, it can be used for forage land (cultivation of perennial grasses), on plots (maps) with sufficient reserves of lowland peat for these purposes after clearing channels and diverting excess water, except for the cultivation of perennial grasses; peat extraction for the production of organic fertilizers (compost) is possible.

Keywords: peatland, monitoring, soil, depleted peat bog, vegetation type

1. Introduction

A peat bog is a complex ecosystem, the main components of which are water, vegetation, and peat. Experts consider the swamp as a group of interconnected biogeocenoses characterized by abundant moisture, specific moisture-loving vegetation, and peat formation [1]. The living conditions of plants here are different from the conditions of forests and meadows. Swamps are characterized by constant or periodic abundant moisture, insufficient aeration, poor nitrogen-mineral nutrition, and constant growth of peat substrate.

There are peat bogs and bogs in almost all natural areas. Grass bogs, for example, are found in all zones of the European part of Russia—from tundra to semi-deserts. Polygonal and bumpy swamps are common in the tundra, upper sphagnum swamps—in the coniferous forest (taiga) zone. The nature of the distribution of bogs, their size, configuration, species composition and structure of vegetation cover, the thickness, and structure of peat deposits are mainly due to climate and geomorphological conditions [1, 2].

The largest areas of peat bogs in the European part of Russia are concentrated in the north and northwest of the coniferous forest (taiga) zone. The dominant position is occupied by convex oligotrophic peatlands, for the development of which the most favorable conditions have been developed here: significant predominance of precipitation over evaporation, rather high relative humidity, proximity to the surface of groundwater and the lack of mineral nutrition in their elements; and flatness of the territory as well as a long history of the development of surface formations. This zone is characterized by intensive peat accumulation and makes up the bulk of Russia's peat reserves [2].

Swamps are also important for maintaining the water level in adjacent biocenoses. Complete drainage of the swamp can ruin the nearby area. If the sea is close, seawater will invade the groundwater used as drinking water in cities located on the coast. Many small rivers, streams, and tributaries of large rivers originate in the upper marshes, and if the marshes are drained, the rivers will lose their sources feeding them. Even when swamps do not share water with rivers, they slow down the surface runoff of water falling to the ground in the form of precipitation, and this is very important, since water should flow down the ground as slowly as possible to prevent erosion. After the campaign to drain the swamps, which was carried out in the past century in the Soviet Union, peat bogs begin to burn every hot summer in the Central Federal District. The main reason for this was the violation of fragile hydrological cycles [3].

In recent years, the marshes have become the object of close attention of scientists. This is not surprising because swamps are not only unique ecological systems but also valuable mineral deposits. The development of swamps is very rapid. The discovery of the richest deposits of oil and gas in the wetlands of Siberia and the Far North, the development of peat, as well as the increase in the area of arable land, all this requires the drainage of swamps. At the same time, there is a threat of their complete destruction. But as a natural landscape, swamps are an integral part of the biosphere. As noted above, they play a major role in the hydrological balance of a number of localities. At the same time, many aspects of the functioning of swamp ecosystems remain unknown until now. Therefore, swamps as a type of plant community require not only comprehensive protection but also fundamental research. Such studies are especially relevant in Russia because in terms of the total area of wetlands, our country ranks first in the world [4].

The peat deposit with its ecologically useful resources is of interest for agricultural production. A peat bog after drainage (reclamation) can be effectively used as a highly productive agricultural land. Peat soils of lowland and transitional bogs surpass chernozems in terms of potential nutrient reserves in a meter layer and, with rational use, are much more productive than sod-podzolic and gray forest soils. As the research results have shown, the highest payback of fertilizers and low cost of high-quality products are achieved on cultivated peat bogs (Уланов).

The peat soils of fens and transitional mires on the potential reserves of nutrients in the m layer are superior to the black soil and the rational use of much more productive sod-podzolic and gray forest soils. Abandoned drained peatlands represent an environmental hazard in connection with a high likelihood of fires, the cause of which is mainly the failure to comply with fire safety in the temporary dry grass, kindling fires, etc. [5, 6].

Drained and abandoned peat bogs pose an environmental hazard due to the occurrence of peat fires, the cause of which is mainly non-compliance with fire safety when burning dry grass, kindling fires, etc. The long-term preservation of the fertility of peat bogs depends on the nature of their use. With incorrect methods of use, rapid mineralization of organic matter occurs, which leads to a reduction in its reserves. Mineralization of organic matter leads to unproductive loss of mobile forms of nitrogen compounds [7, 8].

Shallow-lying and shallow-contoured peat bogs (up to 10 hectares) should be allocated for cultural hayfields and pastures. When developing methods of intensification of agriculture on peat soils at different stages of anthropogenic evolution, an objective assessment of the state of properties and the forecast of their possible changes over time under the influence of anthropogenic and abiotic factors is of utmost importance. The introduction of agricultural land plots with small-contour peat deposits into circulation is of practical interest for land users, which is associated with the fact that these soils are potentially highly fertile and can be successfully used for growing fodder crops. But at the same time, such peat deposits have a feature that is associated both with the specifics of the use of peat soils and with their periodic water logging, since they are mainly located at the edge of the forest and at the edge of fields with mineral soils [9–11].

There are 9260 small-scale (up to 10 ha) peat deposits on the territory of the Russian Federation, which occupy an area of 108.6 thousand hectares in the zero boundary of the deposit [12]. The largest number of shallow-lying and shallow-contoured peat bogs is located in the North-Western, Central, and Volga Federal Districts. So, in the Central Federal District, out of 7287 explored deposits, 2390 are small scale and 1298 are small scale and protected, that is, almost half. On the territory of the Vladimir region, where these studies were conducted, out of 723 peat deposits, 421 are deposits with an area of 1 to 10 hectares, where proven peat reserves in the sum of categories A + B + C1 and C2 (144 deposits) and forecast resources in category P1 (277 deposits) amount to 4277 thousand tons. Small-contour peatlands are often located on the edges of fields with mineral soils and adjacent to forests; their use in agricultural production has its own characteristics and is associated with the characteristics of peat soils. Currently, such deposits are abandoned and overgrown with forests. The degradation of landscapes entails a deterioration in the quality of vegetation and soil cover, water and temperature balance of the peat reserve territory, and soil composition, which provokes the development of water and wind erosion. At the same time, there is a transformation of the forest-meadow agricultural landscape with the dominance of meadow plant species into post-swamp forest-shrub-grass-sedge landscapes with significant participation of secondary forest phytocenoses [10]. In addition to negative changes in vegetation cover and water and temperature balance of the territory, soil degradation develops. With illiterate and irrational exploitation of the peat bog, rapid mineralization of organic matter occurs, which leads to a reduction in its reserves and unproductive loss of nutrients.

Soil physical, chemical, and biological properties collectively determine the quality of the soil. The biological properties of the soil are characterized by the presence in them not only of various microorganisms but also of the processes of plant growth. Dying plants and their parts, deposited in the soil, are enriched with nutrients in forms resistant to leaching. The root system of plants moves minerals from the lower layers to the upper ones. The biological process is, thus, a factor in the concentration of nutrients in the soil. Both mineral salts and synthesized organic substances containing a lot of nitrogen are concentrated in the upper layer [7, 8].

Agrochemical surveys are carried out in order to obtain information about the content of plant nutrition elements in the soil and as a consequence of the level of its fertility. Agrochemical examination allows more rational use of fertilizers and to minimize their negative impact on the environment. As a result, agrochemical cartograms of the content of elements, agrochemical essays, and application maps of fertilizer application are created. We determine the basic properties of the soil, which in one way or another can affect the growth and development of plants. One of the most important indicators determined by agrochemical analysis is the reaction of the soil solution (pH), the content of mobile phosphorus and potassium required by plants [7, 8].

The importance of different plants in soil formation is not the same. Under the forest, if there are no herbaceous plants, organic substances do not accumulate. Due to the constant presence of fulvic acid here, salts are washed out of the upper layer, and the soil formed on the carbonate rock acquires an acid reaction (podzol formation process). Under herbaceous plants, due to their gradual death, organic residues are formed, which accumulate mainly in the thickness of the soil. The reaction of the soil solution here is more often neutral or close to neutral. Against this background, bacteria settle. Under the action of bacteria, the organic remains of plants turn into humus (humus), which gradually accumulates in the soil and improves it (the sod process) [7, 10].

The purpose of our study is to monitor changes and the state of agrochemical and other characteristics of anthropogenically transformed peat soils, depending on the directions of use of the developed peat bog to obtain data used to develop the most promising and rational ways of using the peat bog.

2. Monitoring the properties of an abandoned depleted peat bog

2.1 Materials and methods

The research was carried out at the Baigush peat deposit, located 1.5 km northeast of the village Baigushi (Sudogodsky district, Vladimir region, 56°078111 N, 40°493809E). This territory belongs to the middle peat-swamp region [2], the geomorphological conditions of which are represented by moraine and alluvial landscapes with the presence of pronounced traces of the last glaciation in the form of finite moraine formations that have undergone severe erosion. In 1943, the peat bog was assigned category C2 (assessed)—the field was intended for agricultural use. In 1963–1965, the massif was used for peat extraction for fertilizers. Until 1963, the thickness of the peat layer averaged 109 m^{-1} (maximum 140 m^{-1}) and in 1975, no more than 40–50 m^{-1} cm, so after 1975, the peat began to be used as hay or pasture. According to the Geological Survey of 1977, the type of peat deposits was defined as transitional, closer to lowland peat (A-15%, R-45%) [12]. The total area of the peat massif was 13.8 hectares and peat reserves—30 cubic meters (or 6 thousand tons at 40% humidity). Reclamation (drainage) was carried out in 1985; the drainage basin was a ravine.

From 1986 to 2014, the area of the peat bog was in the land use of the experimental production facilities of the Institute; on a small area of the peat bog (I and II peat charts), where the peat was almost completely worked and which was almost not flooded, grain and fodder crops were cultivated. On the remaining maps, peat was extracted for compost production; peat on maps III, IV, and V was partially worked.

Currently, the territory of the peat bog is completely abandoned. In 1998, a soil-agrochemical survey of the territory of the peat massif was carried out; the layout of peat maps and conventional reference points are shown in **Figure 1** and **Table 1**. Monitoring of changes in some agrochemical properties of the soil on peat charts of the Baigush peat deposit (**Figure 1**) was carried out 20 years after the first survey of the peat massif. The research was carried out by the route expedition method at the same survey points as in 1998 (**Table 1**).

In 2017–2018, to determine the change in the basic agrochemical properties of the fine-contour shallow peat bog, a soil-agrochemical and field geobotanical survey of the peat massif was carried out using the methods [13–16].

Geobotanical description, determination of agrophysical, and biological properties of the soil by survey points were carried out for the first time in 2018. A geobotanical survey was carried out in biogeocenoses of 15 locations within the boundaries of five peat charts, which consisted of determining plant species and their abundance on the Drude scale [17]. Agrochemical parameters of the soil of the object were determined in accordance with state standards, nitrifying ability by the Kravkov method, cellulolytic activity by the application method, density, and density of the solid phase of the soil by the weight method.

2.2 Results and discussion

During the research, an expeditionary geobotanical survey of the peat massif was carried out, during which 80 plant species and their abundance were determined according to the Drude scale in biogeocenoses of 15 conditional reference points (locations) on five peat maps. At the moment, the geolocations of the points are fixed in the coordinate system. The vegetation cover of the surveyed territory is represented

Figure 1.
Layout of peat charts on the Baigush peat deposit: Sudogodsky district, Vladimir region, 56°078111 N, 40°493809E (used app "Google earth").

V chart		Chart canal		Main canal
	*15*1	*14	*13	
IV chart		Chart canal		
	*10	*11	*12	
III chart		Chart canal		
	*9	*8	*7	
II chart		Chart canal		
	*6	*5	*4	
I chart		Chart canal		
	*1	*2	*3	
Dirt road				

Table 1.
Location of peat charts and survey points on the Baigush peat deposit.

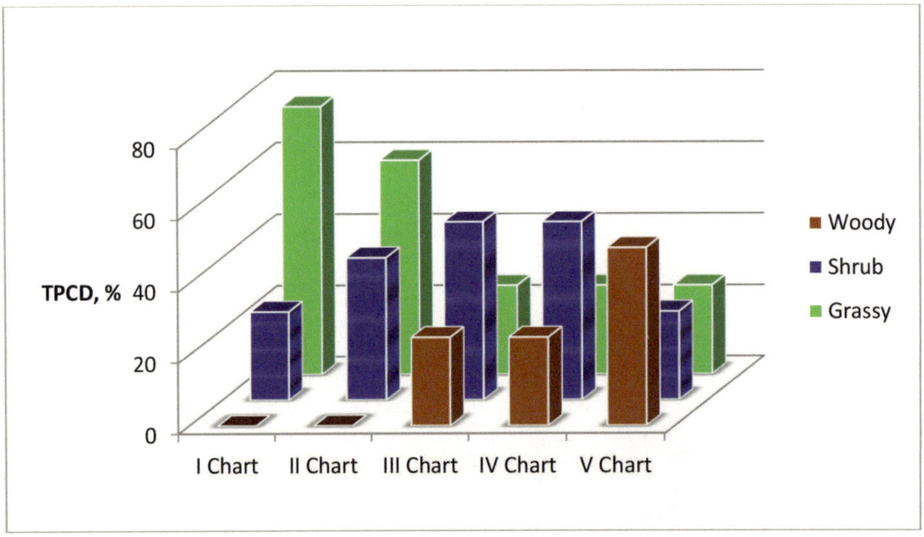

Figure 2.
Vegetation types on peat charts.

by meadow and forest phytocenoses. According to the results of the geobotanical survey of the object, the predominant types and types of vegetation were established (**Figure 2**).

On I chart, the composition of the herbaceous tier is diverse; the total projective cover degree (TPCD) of grasses is >70%: Veronica oakwood (*Veronica vulgaris L.*), ground vane (*Calamagróstis epigéjos L.*), bonfire without a tail (*Bromopsis inermis L.*), clovers, bluegrass, sharp sedge (*Carex acuta L.*), common tansy (*Tonacetum vulgare L.*), fine vole (*Agrostis capillaris L.*), creeping wheatgrass (*Elytrigia repens L.*), meadow timothy (*Phlum pratense L.*), common yarrow (*Achillea millefolium L.*), and horsetails. Shrubby vegetation type (TPCD >20%) is represented by shaggy willow (*Salix lanata L.*) and holly willow (*Salix acutifolia L.*) (**Figure 3**).

Figure 3.
Vegetation on the I chart.

Figure 4.
Vegetation on the II chart.

On II chart, the proportion of herbaceous vegetation has decreased; the TPCD is >60%: field grass (*Cirsium arvense L.*), field cornflower (*Centaurea jacea L.*), common goldenrod (*Solidago virgaurea L.*), bonfire (*B. inermis L.*), clovers, bluegrass, acute sedge (*C. acuta L.*), sedge, thin vole (*A. capillaris L.*), creeping wheatgrass (*E. repens L.*), meadow timothy (*Phlum pratense L.*), and horsetails. The shrubby vegetation type (TPCD ~40%) is represented by shaggy willow (*S. lanata L.*), holly willow (*S. acutifolia L.*), and dog rose (*Rosa canina L.*) (**Figure 4**).

On the territory of III–IV charts, shrubby vegetation type prevails, (TPCD >50%): shaggy willow (*S. lanata L.*), holly willow (*S. acutifolia L.*), and common hazel

Figure 5.
Vegetation on the III chart.

(*Corylus avellana L.*). Variegated grasses (TPCD ~25%) are replenished with moisture-loving vegetation: field grass (*C. arvense L.*), wood angelica (*Angelica sylvestris L.*), common goldenrod (*Solidago virgaurea L.*), clovers, bluegrass, forest cupyr (*Anthriscus sylvestris L.*), acute sedge (*C. acuta L.*), bubble sedge (*Carex vesicaria L.*), tenacious bedstraw (*Galium aparine L.*), vaginal fluff (*Eriophorum vaginatum L.*), and horsetails. The woody type of vegetation (TPCD ~25%) is mainly represented by rhombic alder (*Ansys rhombifolia L.*) and scots pine (*Pinus sylvestris L.*) (**Figures 5** and **6**).

On the territory of V chart, the predominant vegetation type is woody (TPCD >50%): mainly it is hanging birch (*Betula pendula L.*), mountain ash (*Sorbus aucuparia L.*), and common pine (*P. sylvestris L.*). Shrubby vegetation type (TPCD ~25%) is represented mainly by shaggy willow (*S. lanata L.*) and holly willow (*S. acutifolia L.*). Motley grasses (TPCD ~25%): ground weiner (*Calamagróstis epigéjos L.*), bluegrass, sharp sedge (*C. acuta L.*), bubble sedge (*C. vesicaria L.*), tenacious bedstraw (*G. aparine L.*), vaginal fluff (*E. vaginatum L.*), horsetails, sod pike (*Deschampsia cespitosa L.*), and acute sitnik (*Juncus acutus L.*) (**Figure 7**).

Thus, the overgrowth of the surface of the developed peat bog with woody-herbaceous vegetation largely depended on the capacity of the residual peat. On the plots that were completely and heavily processed (peat thickness from 0 to 50 cm) and retired from agricultural use in the mid-90s (point № 8, 9, 10, 15), a forest with its inherent tiering was formed: the bulk of woody vegetation is hanging birch (15–22 m), common mountain ash (2–4 m), and common pine (up to 3 m); shrubs are mainly represented by willows; the herbaceous vegetation is described in detail above (charts III–IV).

With the thickness of the residual peat layer of 70 cm or more, the growth and development of woody vegetation occurred slowly (point № 4, 7, 11, 12, 13, 14). The

Figure 6.
Vegetation on the IV chart.

Figure 7.
Vegetation on the V chart.

multi-tiered nature of the forest is poorly expressed: there are single birches (up to 5–6 m), aspens (2–3 m); shrubs are represented by willows, rose hips, and raspberries. Since these areas are under water until the end of spring, the herbaceous vegetation is mainly represented by moisture-loving plants: ground weinik (*Calamagróstis epigéjos L.*), swamp horsetail (*Equisétum fluviatile*), sharp (*C. acuta L.*) bubbly sedge (*C. vesicaria L.*), and fluff (*Erióphorum vaginátum L.*).

The data of the soil-agrochemical survey of a shallow-contour shallow-lying peat bog on 15 reference points, depending on the cultivation and intensity of the use of peat-boggy soils according to the maps, are presented in **Table 2**. As a result of observations, the change in the content of the main biogenic elements over a 20-year period (from 1998 to 2018) has been established. The content of mobile phosphorus and exchangeable potassium has changed to the greatest extent. So, on the I map, in the soil layer of 0–80 cm, the content of mobile phosphorus and exchangeable potassium changed slightly, which can be explained by the fact that the territory of the map was in agricultural use for a long time. The areas of the remaining charts have been abandoned for more than 20 years; in the spring, they are mostly under water; and in dry years, the territory of the II chart was partially used in the agricultural production.

On map V, an increase in the content of mobile phosphorus was found in soil layers from 0 to 80 m^{-1}, which can be explained in the absence of fertilizers by its biogenic accumulation, since phosphorus, as shown in the studies of T. Kulakovskaya et al., has an extremely weak migration ability, and no more than 3–5% of its total

Depth, cm	Content of mobile phosphorus (P_2O_5), mg/kg			Content of mobile potassium(K_2O), mg/kg			pH		
	1998	2017–2018	Δ	1998	2017–2018	Δ	1998	2017–2018	Δ
I chart									
0–33	48.9	50.6	+1.7	45.7	39.2	−6.5	6.6	6.4	−0.2
33–80	10.1	12.8	+2.7	13.5	14.0	+0.5	6.4	6.35	−0.05
II chart									
0–28	57.3	69.8	+12.4	69.7	62.4	−7.3	6.23	6.3	+0.07
28–80	27.3	35.9	+8.6	51.5	32.8	−18.3	4.7	4.9	+0.02
III chart									
0–26	51.0	55.9	+4.9	62.3	61.2	−1.1	4.5	4.6	+0.1
26–80	15.0	17.4	+2.4	11.0	11.5	+0.5	30	3.5	+0.5
IV chart									
0–34	15.9	27.1	+11.2	76.2	40.8	−35.4	4.5	4.4	−0.1
34–80	12.5	16.9	+4.4	64.0	20.0	−34.0	3.9	4.05	+0.15
V chart									
0–35	17.4	42.8	+25.3	44.1	40.5	−3.6	3.9	3.9	0
35–80	11.5	24.3	+12.8	47.0	25.7	−21.3	3.3	3.6	+0.3

Table 2.
Changes in some agrochemical indicators over a 20-year period (average by charts).

Peat chart	Point	Nitrification capacity, mg kg^{-1} for 30 days	Cellulosolyical activity, %	Soil density (D), g m^{-3}	Poriness, %	The power of the residual peat layer, m^{-1}
I	1	31.0	17.0	1.8	39.0	<10
	2	32.4	17.5	1.40	39.5	<10
	3	32.8	17.5	1.49	26.2	<10
II	4	45.9	17.5	1.54	32.0	<10
	5	49.2	12.5	1.39	36.6	<10
	6	43.7	17.5	1.18	46.1	<10
III	7	14.8	70.0	0.68	65.4	31–50
	8	25.8	10.0	1.21	41.4	<30
	9	35.4	17.5	1.22	47.8	<30
IV	10	13.2	17.5	1.20	34.0	<30
	11	11.1	55.0	0.77	41.0	31–50
	12	11.9	57.5	0.77	59.0	>51
V	13	14.8	77.5	0.72	60.6	>51
	14	14.2	62.5	0.81	57.8	>51
	15	14.9	25.0	0.78	57.4	31–50

Table 3.
Biological and agro-physical soil properties peat.

reserves is washed out [7, 8, 18, 19]. In addition, it was shown that in shallow, weak- and medium-azole peat bogs, with commercial water regime and high groundwater aquifer, they can penetrate into the subsoil and underlying layers [5, 6, 19], which our observations also showed. Unlike mobile phosphorus, an increase in the reserves of exchangeable potassium in the soil of maps IV–V in layers from 0 to 80 m^{-1} was not observed, since its high mobility and intensive use are increasing, especially in the soil layer 30–50 m^{-1}, where the bulk of the roots are located. The pH values in the peat bog soil have not changed much.

The difference in the data on the biological and agrophysical properties of peat soil presented in **Table 3** can be explained by the difference in the degree of cultivation of the studied soils and the residual peat layer on the charts. A direct relationship has been established between the thickness of the residual peat layer and the cellulolytic activity and porosity of the soil; an inverse relationship is established between the thickness of the peat layer and the nitrifying ability and density of the soil.

The nitrifying ability of the upper soil layer decreases with increasing peat thickness (point № 7, 10–15), cellulolytic activity, and porosity; on the contrary, it increased at these points in the soil. With an increase in peat thickness, soil density indices decreased from 1,18-1,8 (point № 1-6, 8-10) to 0,68–0,81 g m^{-3}.

3. Conclusion

During the soil-agrochemical survey of five peat bog maps, a change in the content of mobile phosphorus over a 20-year period was detected, which noticeably increased in the soil layer 0–80 m^{-1} on the fifth map, and the content of exchangeable potassium significantly decreased in the soil on the fifth map and the fourth and fifth cards. During the monitoring of the condition of the developed fine-grained marsh peat, a direct relationship was established between the thickness of the residual peat layer on the maps and the cellulolytic activity and porosity of the soil as well as an inverse relationship between the thickness of the peat layer and the nitrifying ability, soil density. In depleted territories, vegetation is mainly represented by various grasses and shrubs, which can be explained by the rather long use of maps in agricultural production; in medium-developed territories, shrubby-woody vegetation prevails, with a peat layer thickness of more than 30 cm; and sedge and fluff dominate in flooded areas.

Author details

Anisimova Tatiana Yuryevna
All-Russian Research Institute of Organic Fertilizers and Peat—A Branch of the Federal State Budget Scientific Institution «Upper Volga Federal Agrarian Scientific Center», Vladimir Region, Russian Federation

*Address all correspondence to: anistan2009@mail.ru

References

[1] Yu AT, Luchanok LN. Problems of effective use of drained peatlands in Russia and Belarus: Theoretical and practical bases (Проблематика эффективного использования осушенных торфяников в России и Беларуси: теоретические и практические основы). Agrochemistry and Ecology Problems;**4**:78-84. DOI: 10.26105/AE.2018.4.33.017

[2] Tyuremnov SN. Peat Deposits. Moscow: Nedra; 1976. p. 488

[3] Sirin A et al. Issues of re-waterlogging of peatlands in the climate reports of the Russian Federation. Earth. 2021;**10**(11):1200. DOI: 10.3390/land10111200

[4] Minaeva T, Sirin A, Kershaw P, Bragg O. Arctic peatlands: Entry in the directory "alive". In: In the Book of Wetlands: II: Distribution, Description and Conservation. Netherlands: Springer; 2017. pp. 1-15

[5] Abel S, Haberl A, Joosten H. A Decision Support System for Degraded and Abandoned Peatlads Illustrated by the Example if Peatalands of the Russian Federation. Germany: Published by Michael Succow Foundation; 2011. p. 48. Available from: www.succow-stiftung.de/tl_files/pdfs_downloads/Buecher%20und%20Broschueren/DSS-Brochure_final_2012_lowres.pdf [Accessed: December 12, 2020]

[6] McNicol G, Knox SH, Guilderson TP, Baldocchi DD, Silver WL. Where old meets new: An ecosystem study of methanogenesis in a reflooded agricultural peatland. Global Change Biology. 2020;**26**(2):772-785. DOI: 10.1111/gcb.14916

[7] Kulakovskaya TN. Optimization of the Agrochemical System of Soil Nutrition of Plants. Moscow: Agropromizdat; 1990. p. 218

[8] Glazovskaya MA. Geochemistry of Natural and Technogenic Landscapes of the USSR. Moscow: High School; 1988. p. 328

[9] Wahyunto W, Supriatna W, Agus F. Land Use Change and Recommendation for Sustainable Development of Peatland for Agriculture: Case Study at Kubu Raya and Pontianak Districts. West Kalimantan; 2020. pp. 32-40. DOI: 10.21082/ijas.v11n1.2010

[10] Ulanov AN, Shelmenkina HH, Smirnova AV. Environmental aspects of reclamation the disturbed wetland ecosystems (Экологические аспекты рекультивации нарушенных болотных экосистем). Subsoil Use Problems. 2017;**2**(13):56-61. DOI: 10.18454/2313-1586.2017.02.056

[11] Wijedasa LS, Page SE, Evans CD, Osaki M. Time for responsible peatland agriculture. Science. 2016;**354**(6312):562. DOI: 10.1126/science.aal1794

[12] Peat deposits of the Vladimir region as of January 1, 1977. Ministry of Geology of the RSFSR. Moscow: Geoltorfrazvedka Trust; 1978. p. 368

[13] Maksimov AI et al. Research Methods of Bog Ecosystems of the Taiga Zone. Science, Leningrad; 1991. p. 110

[14] Sukachev VN et al. Fundamentals of forest typology and biogeocenology. Science, Moscow. 1972;**V.1**:424

[15] Fedorets NG, Medvedeva MV. Methodology for Studying Soils of Urbanized Territories. Petrozavodsk:

Karelian Scientific Center of the Russian Academy of Sciences; 2009. p. 84

[16] Semenenko NN. Agrochemical Methods for Studying the Composition of Nitrogen, Phosphorus and Potassium Compounds in Peat Soils. Navuka, Minsk: Belarus; 2013. p. 78

[17] Cherepanov SK. Vascular Plants of Russia and Neighboring States (within the former USSR) (Сосудистые растения России и сопредельных государств (в пределах бывшего СССР)). St.Petersburg: Peace and Family; 1995. p. 990

[18] Ulanov AN. Peat and Worked-out Soils of the Southern Taiga of the Euro-North-East of Russia. Kirov: JSC "House of the Press - Vyatka"; 2005. p. 320

[19] Kovalev NG, Pozdnyakov AI, Musekaev DA, Pozdnyakova L. Peat, Peat Soils, Fertilizers. Moscow: VNIIMZ; 1998. p. 239

Chapter 7

Towards Collaborative Cluster Management for Fire-Resilient Peatlands in Indonesia

Johan Kieft

Abstract

Wildfires on peat lands in Indonesia have been a major cause of globalGHG emissions and has had an irreversible impact on the health of millions, in 2020, the goi decided to introduce the so-called fire protection association or so called's which is seen globally as best practice in terms of integrated fire management governance and in Indonesian named clusters. In 2020., a pilot involving three districts in fire prone landscapes introducing fire protection associations was commenced to understand if FPA could be employed in the Indonesian context could deliver similar results, the results and developed approach lead to a decline in fire incidence in the target districts as opposed to the district in the province. Hence the cluster approach indeed proved by better alignment of private and public fire capacity in addition to improved early warning capacity. The results underline that the necessary processes that are gender sensitive and socially inclusive can be adapted to all jurisdictional levels and enable effective collaboration of relevant government agencies. Cluster maintains the core principles of fire protection associations and integrated fire management, in line with international best practices in disaster risk reduction. Furthermore, Changes allow for improved local livelihoods of communities depending on peat lands, as hydrological restoration and reafforestation enables local communities to again rely on swamps for their livelihoods.

Keywords: integrated fire management, peat, haze, governance, fire-resilient peatlands

1. Introduction

The 2015 fire crisis in Indonesia was an economic and environmental disaster. With 2.6 million hectares of land burned, it cost the country an estimated US$16.1 billion (IDR 221 trillion), equivalent to 1.9% of GDP. Smoke pollution also contributed to irreversible impacts on the lives of 100,300 people across Indonesia, Malaysia and Singapore [1], with more than 500,000 cases of acute respiratory infections. Immediate health costs were estimated at US$151 million [2]. Up to 90% of the smoke pollution came from fires on peatlands, which release 3–6 times more particulate matter than fires on other soil types [2].

Quick and effective rewetting and restoration of peatlands are essential to prevent further degradation through wildfire incidence. In response to the 2015 fires, the Indonesian government introduced the concept of peat hydrological units (Regulation

57/2016). In 2016, the government started working through a south-south exchange with South Africa with the support of UNEP to establish clusters of fire protection associations, normally covering a peat hydrological unit [3].

Best practices are emerging in the global literature on integrated fire management in tropical peatlands (e.g. [4]). These include the establishment of fire protection associations and effective collaboration between land users, high levels of public awareness, a holistic and integrated approach, functioning public-private partnerships, government resources and a regional approach that enables resources to be pooled and better matched to threats.

This article reports on a UNEP project supported by the USAID Bureau for Humanitarian Assistance (BHA) initiated 2-year program for 2019–2021 in partnership with Kemitraan and Working on Fire/Kishugu[1] from south Africa and Institut Pertanian Bogor (IPB)—centre for climate risk and opportunity management in Southeast Asia Pacific (CCROM - SEAP), which was extended due to the impact of the covid pandemic. UNEP has had intensive consultation with the Ministry of Environment and Forestry and the Coordinating Ministry for Economic Affairs to support the implementation of the following project two outcomes:

1. Prototype Fire Protection Associations (FPA's)/Klasters are set up and operational and implementing IFM to reduce forest and peatland wildfires in their areas. The initial prototypes planned by the Government of Indonesia are seven Klasters across the country, of the 14 peat fire-prone districts.

2. South-South collaboration and outreach efforts generate momentum for countries and partners to accelerate the use of integrated fire management to reduce forest and peatland wildfires.

3. The design of a fire risk monitoring system can improve collaboration at national level with private sector and BNPB and improve preparedness at an earlier phase. Once operationalized, the Fire Risk System (FRS[2]) will provide a wide range of government actors with probabilistic information on the likelihood of fire activity at the provincial and district level.

2. Landscape selection

As part of the project through the FRS fire vulnerability maps, project locations were identified through an assessment of fire risk and fire vulnerability based on these maps as shown in **Figure 1**. Four categories of variables—biophysical, socioeconomic, exposure and adaptive capacity—were used in this assessment of fire vulnerability. For each of the selected 10 provinces and eight districts, 11 key variables were included. These were: (i) peat depth, (ii) land cover/use, (iii) distance to road, (iv) distance to the river, (v) distance to the village centre, (vi) land system, maps. Land system data is derived from the land system map provided by the Regional Physical Planning Project for Transmigration (RePPProT). For more information, see [5]. (vii) percentage of timber plantation concession area per sub-district, (viii) percentage of

[1] See:https://kishugu.com/working-on-fire/

[2] See: http://kebakaranhutan.or.id/

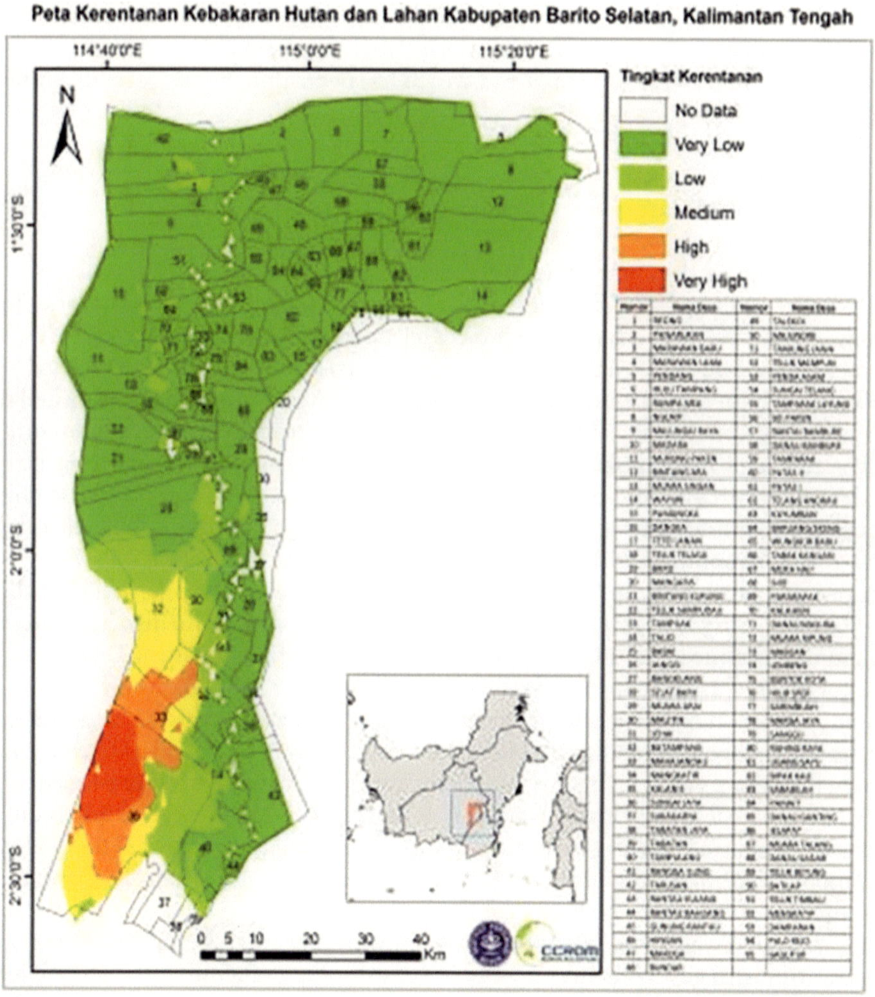

Figure 1.
Fire risk vulnerability map of Barito Selatan (data from 2015). Source: Ipb-ccrom (Bogor agricultural university—Centre for climate risk and opportunity management in Southeast Asia pacific, fire risk monitoring system (see: http://kebakaranhutan.or.id/).

palm oil concession area per sub-district, (ix) percentage of logging concession area per sub-district, (x) population density, and (xi) regional gross domestic product.

The number of satellite-detected fires per km^2 was used as the main measure of spatial and temporal occurrence of fires, using only high confidence locations were applied, with more than 50% certainty of fire activity (based on the official Indonesian fire data), which has data sources from four satellites, namely Terra Aqua, NOAA, SNPP, and Landsat 8, as well as weather data from BMKG. The data in SIPONGI is also more accurate because it contains information about the location at the village level and the status of the land. Vulnerability was calculated from scores and weights of vulnerability indicators, using composite mapping analysis (CMA) [6], resulting in vulnerability maps (e.g. **Figure 1**). The above-described fire risk monitoring system was verified following stakeholder consultation with key land users, mainly smallholders, who had lost perennial crops to fires in previous fire episodes (1997/98, 2002, 2006, 2009 and 2012) [5]. Using fire risk and vulnerability mapping, an area of around 20,000 ha was identified, where during recent years, fires affected more than 100,000 people and which has been emitting close to 90,0000 mt CO2 eq/year.

3. Local collaboration

To sustain impactful, bottom-up water governance structures at the landscape level, it is fundamental to effectively engage land-use managers and communities in damming and rewetting efforts. The project used small grants as incentives, to improve community welfare through the development of horticulture, fisheries and other livelihood activities, paid when people were actively involved in rehabilitation activities such as canal blocking. UNEP-led peatland rehabilitation efforts support community involvement in peatland forest fire control through provision of alternative and sustainable and profitable environmentally-friendly activities. In this way, it is also hoped that targeted communities will desist from illegal logging or slash-and-burn farming.

In close collaboration with the district government and the National Peat Restoration Agency (*Badan Restorasi Gambut*, BRG), dams were constructed in the canal between Sungai Mentangai and Sungai Purun in south Barito district, central Kalimantan Province in 2018, aimed at rewetting higher fire risk areas. Dam construction started as far inland as possible to limit environmental damage by heavy equipment used in construction, following an external environmental assessment. The project constructed compacted peat dams in 2019—2022 and plans to construct them in the next 2 years. Compared to other dam types, these are less expensive (US$500–1000 each), last longer, and have long been used by the plantation industry, with many thousands having been already built in central Kalimantan. The local government has financed most of this work with the project financing the design costs and building smaller structures for secondary channels.

Reforestation was also a key initial part of the project and is being continued by communities with government support, with nyamplung (Calophyllum inophyllum) planted in large numbers during the project, which produces excellent timber, fruit and medicinal honey.

3.1 The cluster approach

Vegetation management and maintenance of stable and correct groundwater levels are both critical to limit fuel availability and prevent peatland fires. Government Regulation 57/2016 recommends maintaining groundwater at no more than 40 cm deep, but ideally near the surface. Effective execution is also required and that considers all local interests. Collaboration between land users enables improved land use planning, specifically regarding drainage, which requires collaboration of land users, which is also required to ensure effective integrated fire management.

The project worked with clusters of fire protection associations to develop arrangements for integrated fire management that were agreed upon with land users and coordinated through incident and command systems. **Figure 2** below shows how policy is guiding the initial piloting, which then should result in national rollout. As UNEP is in the process of both working on the financing and working towards nationwide implementation of the Klaster approach. Within the current project design, UNEP is preparing for a next phase to work towards a nationwide implementation

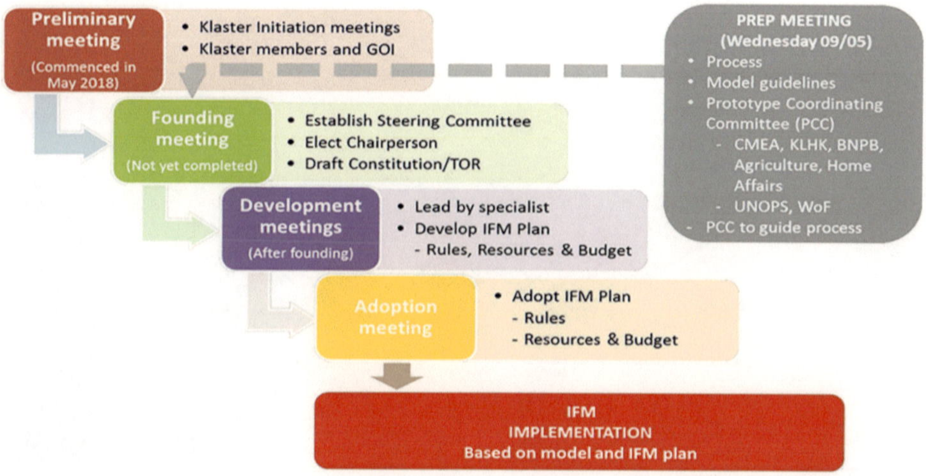

Figure 2.
Flow diagram with key activities for cluster establishment (UNEP, 2020).

of these clusters. At the institution level, currently, Kemitraan as SIAP partners are conducting a study on which model is best suited, either through a so-called special district service agency called BLUD or other forms like a special district government entity or a UPT. The process that has been tested and proved to be effective is presented below (**Figure 2**).

These steps go hand in hand with guidelines for Klaster establishment in line with established GoI legislation and procedures. This includes:

1. Cluster regulation outlines the internal governance

2. Cluster establishment guidelines guiding members in implementing integrated fire management

3. Finetuning of the guidelines for the establishment of Klaster will be done in collaboration withthe BNPB training centres to ensure alignment with existing ICS guidelines, including those of private sector partners. The SIAP project deputy cluster managers will facilitate through joint knowledge-sharing sessions and facilitation to come to agreed ICS-based procedures from the community level up in terms of agreed As such to prevent fires. Improving land use practices and facilitating consensus between cluster members on a land use plan aligned with the required groundwater level for sustainable use of peatland, in turn, create fire-resilient landscapes. Managed by land users through clusters, this also leads to economic benefits. It is important that cluster organizations and members see clear and immediate benefits, as the will of stakeholders is critical to gain momentum. UNEP, in collaboration with a programme in South Africa, conducted a policy benchmarking study. Through stakeholder engagement, exchange visits and joint evaluations, a joint perspective was created for fire protection associations.

The government decided to apply to use such clusters to improve collaborative landscape management, particularly in peatlands. Using fire protection associations

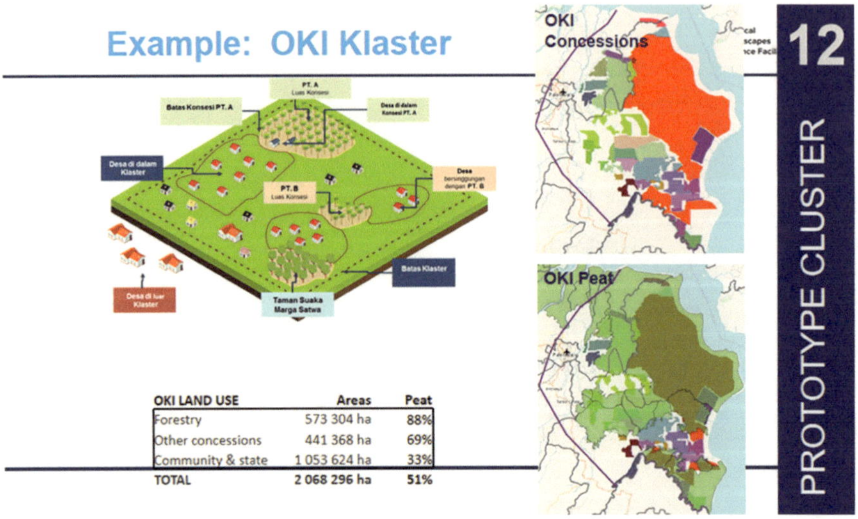

Figure 3.
An example of a cluster for Oki district, South Sumatra (Source: UNEP, 2018).

as a basis for peatland management also provides the necessary scale and resources, as a participatory mechanism for preventing wildfires. Currently, the project is based on pilots to support the development of guidelines (**Figure 3**).

4. Results

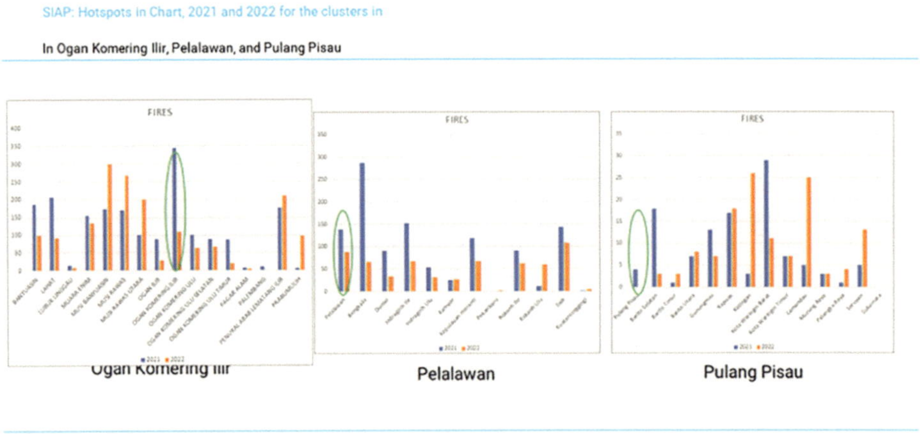

Supported by UNEP, the SIAP Project implementing partners (Kemitraan and CCROM IPB) organized two-phased training on how to use the FRS application for early risk detection and in combination with the cluster members' capacity better aligned through standard operating procedures at cluster levels increased efficiency in fire suppression was achieved. The training conducted in Pulang Pisau was attended by 33 people representing the military, police, FMU/cluster companies, community-based fire brigade and local government offices, while 12 companies attended the training in Pelalawan in addition to the local disaster management agency, fire department and government offices. In Ogan Komering Ilir, the training trained 24 people from local government offices, Klaster-affiliated companies, the military and the police to increase their capacity to determine fire risks and plan basic preventive measures. In both cases, cluster worked towards increasing alignment in terms of land use planning.

4.1 Clear benefits

This shows that fire-resilient landscapes can be realized by including water management as an essential element of fire prevention, supported by clusters, and aligning land and forest use planning across management units, districts and communities. Improved water management has impacts on reducing greenhouse gas emissions from peat decomposition and subsidence but improving land use adjusted to drainage depth also requires a reduction in fires [7].

Based on these experiences, UNEP and its partners have developed clear procedural guidelines on how to establish clusters as described above. These detail the necessary processes that are gender sensitive and socially inclusive, can be adapted to all jurisdictional levels and enable effective collaboration of relevant government agencies. Cluster maintains the core principles of fire protection associations and integrated fire management, in line with international best practices in disaster risk reduction.

Furthermore, changes allow for improved local livelihoods, as hydrological reafforestation enables local communities to again rely on swamps for fish (kerapu and others) and products from native tree species such as sago (*Metroxylon sago*), jelutung (*Dyera polyph*) and gemor (*Nothaphoebe coriacea*). Other benefits are reduced fire incidence, and subsidence leading to subsidence [8] that improve overall human well-being, including for other land users, in particular indigenous communities through empowering of indigenous institutions like *Handils, which* are indigenous land-use systems, as practices in the cases of barioto Selatan. And as well as in some areas in Sumatra [9]. The term *handil* refers to the hand-dug, man-made waterways to gain access to farming fields in these areas as well as to the associations that manage the natural resources of the *handil* area, consisting of the *handil* canals and the surrounding agricultural land.

The cost implications based on a financial assessment [5] suggested that adopting the cluster approach would allow the government of Indonesia to make significant fiscal savings. However, a more in-depth study of actual expenditure on wildfire prevention and suppression between all agencies and departments, including at the provincial and local levels, would provide further insights.

4.2 Next steps

UNEP is preparing for a second phase of nationwide implementation of these clusters, and Kemitraan is conducting a study on which model is best suited, either through a special district service or district government agency, and ideally including indigenous institutions like Handils as members. As such, institutions in Kalimantan have similarities with Dutch water boards and are generally recognized as good managers of collective natural resources. They are relatively autonomous, effectively managing their area and its waterways, and have a form of democratic governance to guard members' interests. They, therefore, have potential to function as institutions for regional, peat dome-based water management, similar to water boards. The social assessment of the project also recognized that the Handil model could be adapted as a peatland conservation management framework. Strengthening such institutions to cover water governance and community-based land use jurisdictions can ensure the sustainable use of peatlands through meaningful community engagement.

In addition. Water boards should be considered as an entry point for improved water governance. There are existing institutional structures in Indonesia, such as Handils, that can facilitate improved water governance. Handils are indigenous land use systems, as practised in parts of Central and South Kalimantan, and Sumatra [9]. Such water boards would ensure sustainability and reduce the chances of leakage through poor governance and lay the groundwork for fire-resilient landscapes addressing both subsidence and emission of GHG emissions [7, 8].

5. Conclusions

This case shows that the use of fire vulnerability as a tool for REDD+ activity selection on peatland can enable local policymakers and planners and reduces fire incidence and can hence deliver tangible greenhouse gas emission reduction and significant livelihood co-benefits. It also lays the foundation for community-driven sustainable development. As water channels are dammed in line with Dohoong et al., [10], the project has been paying out small grants to improve community welfare through the development of horticulture, fisheries and other livelihood opportunities. More recently, the government of Indonesia has also been providing village development grants to communities. In return for grants, people are obliged to be actively involved in peatland restoration. The project also trained four community-based fire brigades in Dusun Hilir that are now able to protect re-vegetated peatland, which has led to good results in the area. The results of rewetting and re-vegetation show that the fire risk system developed by the project allows for improved targeting of ecosystem restoration activities and so reduces the impact of smoke pollution that has affected tens of thousands of people in the last few years [1, 2].

To, significantly, reduce fire risk in Indonesian peatlands requires the establishment of land user associations in hydrologically defined areas [11]. These should be supported with the use of risk-based mapping tools to produce drainage-based land use plans that include forest, non-forest and community land uses. Communities must agree on joint planning objectives regarding rehabilitation, restoring peatlands through hydrological restoration (by raising groundwater level), and rehabilitating peatlands with paludiculture crops. In this way, and building on indigenous practices, fire-resilient landscapes can be co-created, and the cluster approach has proved to be a useful institutional vehicle for collaborative peatland management in particular against a baseline of increased risk due to climate change [12].

Author details

Johan Kieft
Land Use and Green Economy, UN Environment Programme, Indonesia

*Address all correspondence to: johan.kieft@un.org

References

[1] Koplitz SN, Mickley LJ, Marlier ME, Buonocore JJ, Kim PS, Liu T, et al. Public health impacts of the severe haze in equatorial Asia in September–October 2015: Demonstration of a new framework for informing fire management strategies to reduce downwind smoke exposure. Environmental Research Letters. 2016;**11**:094023

[2] Glauber AJ, Moyer S, Adriani M, Gunawan I, Mileva E, Harimurti P, et al. The Costs of Fires: An Economic Analysis of Indonesia's 2015 Fire Crisis. World Bank, Jakarta. 2015. Available from: https://openknowledge.worldbank.org/handle/10986/23840#:~:text=In%20a%20five%2Dmonth%20period,no%20means%20a%20singular%20event

[3] Kieft, J, van Oosthuizen N. BUKTI - A framework for Integrated Fire Management in the fire prone peat districts of Indonesia, unep/kishugo group, internal document. 2016

[4] Kieft J, van Oosthuizen N, Wison t. Fire protection associations as a collaborative mechanism for a landscape-based approach to integrated fire management for reducing haze and greenhouse gas emissions in Indonesia. Biodiversidade Brasileira-BioBrasil. 2019;**1**:292-294

[5] Kieft J, Smith T, Someshwar S and Boer R. Towards anticipatory management of peat fires to enhance local resilience and reduce natural capital depletion. In: Renaud F, Sudmeier-Rieux K, Estrella M, Nehren U, editors. Ecosystem-Based Disaster Risk Reduction and Adaptation in Practice. Advances in Natural and Technological Hazards Research. Berlin, Germany: M.N.U. Springer; 2016;**42**:361-377. DOI: 10.1007/978-3-319-43633-3_16. ISBN 978-3-319-43633-3

[6] Hoha AS, Siddik A, Saharjo BH, Boer R, Ardiansyah M. Spatiotemporal distribution of peatland fires in Kapuas district, Central Kalimantan province, Indonesia. Agriculture, Forestry and Fisheries. 2014;**3**(3):163-170

[7] Wösten JHM, Clymans E, Page SE, Rieley JO, Limin SH. Peat–water interrelationships in a tropical peatland ecosystem in Southeast Asia. Catena. 2008;**73**(2):212-224. DOI: 10.1016/j.catena.2007.07.010

[8] Hooijer A, Page S, Jauhiainen J, Lee WA, Lu XX, Idris A, et al. Subsidence and carbon loss in drained tropical peatlands. Biogeosciences. 2012;**9**:1053-1071. DOI: 10.5194/bg-9-1053-2012

[9] Lubis ZB. Social mapping of access to peat swamp forest and peatland resources. In: Working Paper. Indonesia: Kalimantan Forest and Climate Partnership (KFCP); 2013. Available from: https://iopscience.iop.org/article/10.1088/1755-1315/58/1/012030

[10] Dohong A, Cassiophea L, Sutikno S, Triadi BL, Wirada F, Rengganis P, et al. Community-Based Peat Rewetting Infrastructure Construction with Canal Blocking. Training Module. Jakarta, Indonesia: Peatland Restoration Agency; 2017

[11] Miettinen J, Hooijer A, Vernimmen R, Liew SC, Page SE. From carbon sink to carbon source: Extensive peat oxidation in insular Southeast Asia since 1990. Environmental Research Letters. 2017;**12**:024014. Available from: https://iopscience.iop.org/

article/10.1088/1748-9326/aa5b6f/meta [assessed september 23 2022]

[12] Yadmiko SD, Murdiyarso D, Faqih A. Climate changes projection for land and forest fire risk assessment in West Kalimantan. IOP Conf. Series: Earth and Environmental Science. 2017;**58**:8. Available from: https://iopscience.iop.org/article/10.1088/1755-1315/58/1/012030